HEAT AND
THERMODYNAMICS

Titles in Greenwood Guides to Great Ideas in Science

Brian Baigrie, Series Editor

HEAT AND THERMODYNAMICS

A Historical Perspective

Christopher J. T. Lewis

Greenwood Guides to Great Ideas in Science
Brian Baigrie, Series Editor

GREENWOOD PRESS
Westport, Connecticut • London

Library of Congress Cataloging-in-Publication Data

Lewis, Christopher J. T. (John Tudor), 1948–
   Heat and thermodynamics : a historical perspective / by
Christopher J.T. Lewis.
         p. cm. — (Greenwood guides to great ideas in science,
ISSN 1559–5374)
   Includes bibliographical references and index.
   ISBN-13: 978–0–313–33332–3 (alk. paper)
   ISBN-10: 0–313–33332–7 (alk. paper)
1. Heat.   2. Thermodynamics.   I. Title.
   QC254.2.L49   2007
   536—dc22   2007018366

British Library Cataloguing in Publication Data is available.

Library of Congress Catalog Card Number: 2007018366
ISBN-13: 978–0–313–33332–3
ISBN-10: 0–313–33332–7
ISSN: 1559–5374

First published in 2007

Greenwood Press, 88 Post Road West, Westport, CT 06881
An imprint of Greenwood Publishing Group, Inc.
www.greenwood.com

Printed in the United States of America

The paper used in this book complies with the
Permanent Paper Standard issued by the National
Information Standards Organization (Z39.48–1984).

10   9   8   7   6   5   4   3   2   1

For Sally Stockley, 1944–2003

CONTENTS

LIST OF ILLUSTRATIONS

SERIES FOREWORD

The volumes in this series are devoted to concepts that are fundamental to different branches of the natural sciences—the gene, the quantum, geological cycles, planetary motion, evolution, the cosmos, and forces in nature, to name just a few. Although these volumes focus on the historical development of scientific ideas, the underlying hope of this series is that the reader will gain a deeper understanding of the process and spirit of scientific practice. In particular, in an age in which students and the public have been caught up in debates about controversial scientific ideas, it is hoped that readers of these volumes will better appreciate the provisional character of scientific truths by discovering the manner in which these truths were established.

The history of science as a distinctive field of inquiry can be traced to the early seventeenth century when scientists began to compose histories of their own fields. As early as 1601, the astronomer and mathematician Johannes Kepler composed a rich account of the use of hypotheses in astronomy. During the ensuing three centuries, these histories were increasingly integrated into elementary textbooks, the chief purpose of which was to pinpoint the dates of discoveries as a way of stamping out all too frequent propriety disputes and to highlight the errors of predecessors and contemporaries. Indeed, historical introductions in scientific textbooks continued to be common well into the twentieth century. Scientists also increasingly wrote histories of their disciplines—separate from those that appeared in textbooks—to explain to a broad popular audience the basic concepts of their science.

The history of science remained under the auspices of scientists until the establishment of the field as a distinct professional activity in the middle of the twentieth century. As academic historians assumed control of history of science writing, they expended enormous energies in the attempt to forge a distinct and autonomous discipline. The result of this struggle to position the history of science as an intellectual endeavor that was valuable in its own right, and not merely in consequence of its ties to science, was that historical studies of the natural sciences were no longer composed with an eye toward educating a wide audience that included nonscientists but instead were composed with the aim of being consumed by other professional historians of science. And as historical breadth was sacrificed for technical detail, the literature became increasingly daunting in its technical detail. While this scholarly work increased our understanding of the nature of science, the technical demands imposed on the reader had the unfortunate consequence of leaving behind the general reader.

As Series Editor, my ambition for these volumes is that they will combine the best of these two types of writing about the history of science. In step with the general introductions that we associate with historical writing by scientists, the purpose of these volumes is educational—they have been authored with the aim of making these concepts accessible to students—high school, college, and university—and to the general public. However, the scholars who have written these volumes are not only able to impart genuine enthusiasm for the science discussed in the volumes of this series, they can use the research and analytic skills that are the staples of any professional historian and philosopher of science to trace the development of these fundamental concepts. My hope is that a reader of these volumes will share some of the excitement of these scholars—for both science and its history.

Brian Baigrie
University of Toronto
Series Editor

PREFACE

Heat and cold and ice and fire have always been unavoidable and powerful aspects of mankind's experience. Speculation about the nature of heat and fire (and their opposites) has long played an important part in man's attempt to understand and control the natural world. This book charts the historical development of modern man's understanding of heat and fire, from the recovery of ancient learning at the end of the Middle Ages up to the beginning of the twentieth century. Over the centuries ideas about heat have been very varied. For medieval philosophers fire was one of the four elements of the earthly creation, along with earth, air, and water. The radical new "mechanical philosophy" of the seventeenth century insisted instead that the sensations of heat and cold were just the result of the more or less vigorous motion of microscopic particles of matter. Most scientists by the turn of the eighteenth century, however, had come to regard heat as a weightless fluid called "caloric," which could flow from a hot body to a cooler one.

"Thermodynamics," the distinctive modern theory of heat and power, emerged during the mid- and late nineteenth century. Thermodynamics sees heat (and all other physical phenomena) as a manifestation of a more fundamental reality, namely "energy." Thermodynamics is one of the core disciplines of modern physical science, and the laws of thermodynamics provide a framework that regulates every other discipline, from chemistry to cosmology. The thermodynamic concept of energy defines modern physics, or modern science even. The article on "Energy" in the *Encyclopaedia Britannica* (9th ed., 1875), asserted that "a complete account of our knowledge of energy and its transformations would require an exhaustive treatise on every branch of physical science, for natural philosophy is simply the science of energy." The First Law of Thermodynamics, the law of the Conservation of Energy, that "Energy can be neither created nor destroyed," is perhaps the most basic axiom, an article of faith almost, of modern science. The Second Law of Thermodynamics, which dictates the direction of all natural processes, may even define the direction of time itself.

This history is intended for students with little or no previous scientific experience. It is told largely through the lives and writings of the philosophers, scientists, and engineers responsible—men such as Galileo, Watt, and Einstein. Their work is set against a wider background of political, economic, institutional, and cultural history; from the Renaissance, through the French Revolution, the Industrial Revolutions, and the growth of world trade. The development of thermodynamics, in particular, is inseparable from the development of the steam engine, which powered nineteenth-century industry and trade. More subtly, many of the pioneers of thermodynamics were sincerely religious men whose belief in a directed, purposeful creation conditioned their scientific thinking. As ever, they all had to struggle for recognition (and funding) from the scientific institutions of their day. This history of ideas about heat can contribute to introducing some of the basic concepts—such as temperature, latent heat, energy, and entropy. It can also serve to show more generally how scientists work, both within the laboratory and as part of society.

In the writing of this book I have been very much supported and assisted by numerous friends and colleagues. I would like to express my special thanks to Richard Noakes, Margaret Pelling, Tim Penton, Jason Rampelt, and Francis Lucian Reid, each of whom kindly read and helpfully commented on various chapters in the course of their writing. I would also like to thank Professor Robert Fox for kindly allowing me to quote extensively from his translation of Sadi Carnot.

INTRODUCTION

All things are an exchange for fire, and fire for all things, even as wares for
gold, and gold for wares.

—Heraclitus of Ephesus

The ancient Greek philosopher Heraclitus of Ephesus (c.550–475 B.C.) believed that fire was the fundamental element from which all else was composed. The nature of heat and especially fire had no doubt long intrigued men (and women). According to ancient Greek mythology, the secret of creating fire had been passed to man by the titan Prometheus, an act of industrial espionage for which he was viciously punished by the gods on Mount Olympus. Rational, philosophical—and recorded—analysis of the natural world can most clearly be traced to Heraclitus and other early Greek philosophers, however. They initiated a tradition of speculative thought that spans two and a half thousand years to the present day. Their ideas were summarized and selectively synthesized by Aristotle, and thence transmitted to the Arabs and the medieval Latin West, where they were integrated with Christian theology. With the so-called Renaissance of learning in the fifteenth and sixteenth centuries, European scholars recovered more systematically the full range of surviving classical philosophical and scientific speculation. It would be quite possible and fascinating to chronicle at length ancient, medieval and, indeed, non-Western ideas about heat and fire.

This book, however, concentrates mainly upon the creation of the modern science of heat, or "thermodynamics," during the mid- and late nineteenth century, the heyday of Victorian industrialization. Historically, however, it is impossible to understand the origins of thermodynamics without tracing the development of ideas and techniques back into the eighteenth century and beyond. This account therefore begins towards the end of the Renaissance, at the turn of the sixteenth century. The now more systematically recovered learning of the ancients, combined with the lively speculations of medieval scholars, provided a rich mixture of ideas that fueled the new natural philosophies and the so-called "scientific revolution" of the seventeenth century (see chapter 1). In many ways, however, the modern quantitative science of heat began with the development of the "caloric" theory—the idea that heat was an indestructible, material (but weightless) "subtle fluid"—at the turn of the eighteenth century (see chapter 2). Although the caloric theory itself faded, from 1815 onwards the theory of heat continued to develop in intimate interaction with the ever-growing industrial importance of the steam engine. The seminal theories of Sadi Carnot in the 1820s (see chapter 3), however, and the painstaking experimental measurements of Joule in the 1840s (see chapter 4) were both initially ignored. The enthusiastic recognition and reconciliation of their apparently conflicting ideas and the early development from 1850 onwards of a coherent mathematical theory of thermodynamics were due to the Scottish scientist William Thomson (later Lord Kelvin) and the German physicist Rudolf Clausius (see chapter 5). The simultaneous development, primarily by Thomson, of the fundamental physical concept of "energy" served to unify, even to create, the modern discipline of "physics." Subsequent elaboration of "dynamical" theories of heat, and specifically the kinetic theory of gases, increasingly reduced heat to a branch of mechanics culminating, at the start of the twentieth century, in an essentially statistical approach to heat and thermodynamics (see chapter 6). (We mentioned above the "Victorian" period; Queen Victoria ruled Britain and the British Empire from 1837 until 1901. It might be claimed, without too much violence to the facts, that the serious development of thermodynamics began in 1837, with Clapeyron's renewed publication of Carnot's neglected ideas, and was completed in 1902 with Gibb's *Elementary principles in statistical mechanics,* and thus coincided almost exactly with her reign.) There remained some phenomena—such as "radiant" heat and the heats of chemical reaction—that did not immediately fit within the new mechanical framework. Our last chapter traces how understanding of these topics developed around the end of the nineteenth century, creating quantum theory and chemical thermodynamics in the process. The last chapter also chronicles the rapid growth of low-temperature physics or "cryogenics."

This series is entitled Great Ideas in Science, and the following history does indeed focus mainly on scientific ideas and the arguments and evidence used to support them. But scientific ideas do not grow in splendid isolation. I have tried, therefore, to present the development of ideas within their wider personal, institutional, economic, technological, and religious contexts. Alongside the interplay of theories and experiments, I have tried to chart personal interests and ambitions, alternative and often rival institutional and national agendas, varied religious and political commitments, and most especially technological and commercial pressures. Wherever possible, I have tried to give an immediate flavor of the concerns and personalities of different scientists by direct quotation from their works and letters.

Much of the substance of science, increasingly through the nineteenth century, is complex, technical, and often *mathematical,* and thus potentially difficult for the uninitiated to understand. I have tried to present the concepts involved in the history of heat and thermodynamics mainly in qualitative terms, and the story will, I hope, be meaningful to those with little or no mathematical experience. Nevertheless, I have been unable to resist the inclusion of a sprinkling of mathematical formulas wherever they seem especially crucial or accessible. Besides, I find that formulas have their own visual, almost magical, fascination, even if only partially understood.

The history of heat and thermodynamics, and especially the origin of the concept of the conservation of energy, has often been controversial. The style of historical study has shifted significantly in recent years. Earlier studies tended to concentrate upon the *internal* history of science, the positive, conceptual contributions to modern science and, to some extent, upon achievements within the British scientific community. The scope of scholarship in recent decades has broadened considerably. The year 1971 was a very good one. Cardwell's *From Watt to Clausius,* subtitled "The rise of thermodynamics in the early industrial age," placed thinking about heat squarely within the context of contemporary power technology (both steam and water) and upon a broad European stage. At the same time, Fox (1971) provided a detailed account, both conceptual and institutional and cultural, of the rise and fall of the "caloric" theory of heat, a theory hitherto neglected, partly perhaps because of its perception as the "failed" opponent of the modern dynamical theory of heat. Subsequently, the development of the dynamical theory of heat, and especially of the kinetic theory of gases, has been the subject of very detailed study by Brush (1976); this study, predominantly conceptual in focus, but enlivened with a wealth of biographical detail, casts great light on the later nineteenth-century development of thermodynamics in general. The historical importance of the caloric theory was confirmed by Fox's edition of the crucial work of Sadi Carnot (1986 [1824]).

A fascinating essay by Kuhn (1977 [1959]) re-ignited a vigorous discussion about the origins of the conservation of energy, with particular reference to the phenomenon of "simultaneous discovery"; Kuhn's approach is largely conceptual, but extremely wide-ranging and subtle. A radical shift to a more systematically institutional and cultural approach to the history of energy appeared in the work of Wise and Smith in the 1990s; Smith's (1998) fascinating, detailed, and sympathetic "cultural history of energy physics in Victorian Britain" is sensitive to tangled webs of intellectual, technical, commercial, political, and religious influences and commitments, and to the pressures of personal career development and disciplinary empire building. Despite these varied and profound contributions, there remain conspicuous gaps in the overall history of heat. It is surprising, for example, that such important figures as Regnault and Clausius remain relatively unstudied. There exists therefore considerable scope for further study.

A word is needed about the labeling of people who practiced science in the early modern period. The word "science," stemming from the Latin *scientia,* is ancient and originally meant "an organized body of knowledge," without the usual modern restriction to knowledge of the material world; in the Middle Ages theology was "the queen of the sciences." On the other hand, the terms "scientist" and likewise "physicist" were only coined in the 1830s, and the latter was not widely adopted until the late nineteenth century. To use the term "scientist" of a seventeenth- or eighteenth-century natural philosopher or mathematician is not just an unimportant anachronism, however; it can obscure shifting institutional and disciplinary boundaries and agendas. Early modern "natural philosophy" was arguably more comprehensive and (ultimately) theological in its aims than the specialized, secular enterprise of modern science; the latter only emerged in the early nineteenth century in the aftermath of the French Revolution. Many recent historians of science have consequently been rightly cautious about the use of essentially nineteenth-century labels such as "scientist" to describe investigators of nature working in the seventeenth or eighteenth centuries. On balance, however, it seems artificial and distracting to exclude the use of "scientist" in preference for such labels of the period as "natural philosopher," "gentleman virtuoso," or "mathematical practitioner," or for some such neutral phrase as "producer of natural knowledge." Thus, although I have usually preferred the appropriate period labels, "scientist" sometimes seems to be the most convenient label for someone investigating natural phenomena in the early modern period, even if the agenda of the enterprise has changed profoundly between the seventeenth and twentieth centuries.

To talk anachronistically of a "physicist" can be even more misleading, however, if we imagine thereby an exponent of the distinctive combination of mathematical theorizing and precise, quantitative experimentation that only gradually emerged through the nineteenth century, definitively in the last quarter. The story of this book has, in fact, very much to do with the emergence of that new discipline, given coherence by the new universal concept of "energy" (see chapter 5). To add extra confusion, the eighteenth-century labels "physician" and "physiologist" were largely equivalent to the term "natural philosopher," all denoting students of "physis" in Aristotle's sense of the "natures" or causes of material things. I have therefore generally resisted the use of "physicist" before the mid-nineteenth century.

Similar issues arise with the use of "amateur" and "professional" before the growth of established professional career structures for scientists and physicists during the nineteenth century. Was the skilled and dedicated experimenter Joule an "amateur" just because he was independently wealthy and funded his own research? Were fellows of the Royal Society necessarily "professional" scientists? In the absence of modern career structures the distinction is potentially very misleading. Nevertheless, rather than use such revealing but distracting phrases as "gentleman specialist," I have generally stuck with "amateur" to denote someone who funded their own scientific activities from some independent source; this is not intended to convey any of the modern implications of a casual, untrained, "amateurish" practitioner.

Inevitably much has been omitted from this history of heat and thermodynamics, even within the limited time span chosen. Obviously much has been omitted in terms of detail. But there are also more systematic omissions. Apart from some discussion of the early development of thermometry, for example, I have devoted relatively little space to the development of instrumental and experimental technique; this is regrettable, because the best scientific ideas are of limited use without experimental confirmation or refutation. I have also omitted much discussion of fire and heat in early alchemy and chemistry, despite the connections with later eighteenth-century concepts of heat. Similarly, apart from the Aristotelian cosmos, the meteorological and physiological functions of heat and fire are only mentioned in passing. Nevertheless I hope that this story will provide an interesting introduction to some of the key ideas of the modern science of heat and thermodynamics, and to the history of those ideas. I hope too that it will provide insights into how scientists work, individually, collectively, and in relation to their wider societies.

HEAT IN THE EARLY MODERN ERA

INTRODUCTION: THE LORD CHANCELLOR'S LAST EXPERIMENT

Francis Bacon (1561–1626), a lawyer and sometime lord chancellor of England, was widely regarded in the later seventeenth century as the founding father and patron saint of modern English science because of his vigorous advocacy of an experimental approach to the study of nature. The manner of his death, as recorded by John Aubrey (1626–97), one of the original fellows of the Royal Society, was symbolic, if ironic:

> The cause of his Lordship's death was trying an Experiment; viz. as he was taking the aire in a Coach...towards High-gate [in north London], snow lay on the ground, and it came into my Lord's thoughts, why flesh might not be preserved in snow, as in Salt. They were resolved they would try the Experiment presently [immediately]. They alighted out of the Coach and went into a poore woman's house at the bottom of Highgate hill, and bought a Hen, and made the woman exentereate [gut] it, and then stuffed the body with Snow, and my Lord did help to doe it himselfe. The Snow so chilled him that he immediately fell so extremely ill...that in 2 or 3 dayes...he dyed of Suffocation. (Aubrey, 1972, p. 179)

Bacon was one of the first thinkers to insist at the start of the seventeenth century that the whole structure of traditional science or "natural philosophy" needed to be demolished and rebuilt on new foundations, using new methods. Bacon emphasized the importance of practical and experimental investigation of nature; other contemporary radicals highlighted instead the value of mathematics and of mechanism in understanding nature. In the course of the seventeenth century these various alternative approaches resulted in a range of new theories and techniques for investigating the phenomena of heat (and cold).

Nevertheless, only in the following century did a coherent quantitative and experimental body of knowledge concerning heat finally emerge, crystallized in the "caloric" theory. But whatever their pretensions to the thoroughgoing reconstruction of knowledge, the early seventeenth-century radicals were still heavily indebted to the classical and medieval heritage against which they reacted.

FIRE AND ANGER: THE CLASSICAL AND MEDIEVAL HERITAGE

Man is born unto trouble, as the sparks fly upwards.

—Job 5:7

Aristotle in the Universities

The dominant intellectual institutions of the medieval period were the universities, which emerged in the late twelfth century and survive into the present day. In general the teaching in these "schools"—whence the later label "scholastic"—was mainly concerned with the three higher "faculties" of theology, law, and medicine. An essential preparation for entry into the higher faculties was provided by the faculty of "arts," which offered a general grounding in Latin grammar, logic, and the techniques of disputation, but also in "natural philosophy," i.e., the philosophical understanding of the natural world. The arts curriculum came to be based very largely on the works of the great Greek philosopher and tutor to Alexander the Great, Aristotle of Stagira (384–322 B.C.); recovered in fairly complete form mainly from Arabic sources during the twelfth and thirteenth centuries, Aristotle's writings had the virtue of being thorough and systematic. Such was Aristotle's reputation among the medieval schoolmen that he was often known simply as "the Philosopher," or as "the Master of Those that Know."

Although developed with great philosophical sophistication, Aristotle's image of the material world was attractively simple, combining common meteorological experience with elementary astronomy. Aristotle's universe was eternal in time but finite in space; the spherical Earth—no educated medieval scholar believed that the Earth was flat—was fixed and motionless at its center and was encapsulated by the celestial spheres of the stars, which revolved around the Earth once every 24 hours. Whereas the celestial spheres were made of an eternal and incorruptible "quintessence," the earthly or terrestrial realm—subject to change, growth, and decay—was composed of the four strikingly meteorological elements: earth, water, air, and fire. These elements were defined by two pairs of opposed qualities, namely, hot and cold and wet and dry: Earth was cold and dry, water cold and wet; air was hot and wet, fire hot and dry. Although the terrestrial elements were also roughly disposed in concentric spheres of water, air, and fire surrounding earth at the center, they were capable of combining with each other in innumerable complex ways. Indeed, they were capable of changing into one another through the interaction of their

Figure 1.1: The medieval, Aristotelian view of a finite, spherical, Earth-centered cosmos. The unchanging celestial spheres of the moon, the sun, and the other planets enclose the ever-changing earthly domain of the four elements: earth, water, air, and fire. From Petrus Apianus (1495–1552), *Cosmographia* (Antwerp, 1539); by permission of the Syndics of Cambridge University Library.

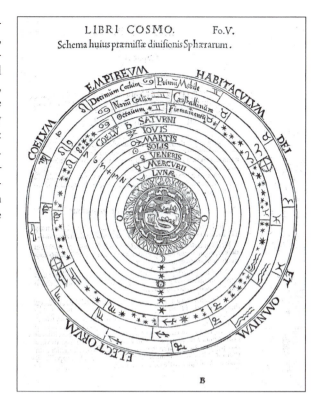

qualities: When water boiled, hot, dry fire combined with cold, wet water to generate hot and wet air. For Aristotle and most medieval scholastics, heat (and cold) were primary qualities, fundamental, irreducible aspects of the material world.

It was a common medieval thought that the structure of the world as a whole (the "macrocosm") was reflected in the structure of the human body (the "little world" or "microcosm"). In particular, the four meteorological elements were mirrored by four physiological "humors"—blood, phlegm, yellow bile (or choler), and black bile (melancholy)—whose balance and interactions determined the character and health of an individual. The humors shared the qualities of the elements: Blood (which corresponded to air) was hot and wet, phlegm cold and wet like water, and so on. The intensity of a given humor (and its associated qualities) in any individual could vary, depending upon their personal balance of humors—their constitutional "complexion" or "temperament"; a constitutional preponderance of blood, for example, would result in a happy, "sanguine" (i.e., "bloody") temperament. In learned medicine, therefore, it was customary to assign degrees of "hotness," "dryness," and so forth, to a patient; this was often done on a scale of one to four, the appropriate degree assessed by appearance and touch. It was mainly from this *medical* context, as a measure of a patient's temperament or "temperature," that the idea of a numerical scale of degrees of heat first emerged.

Renaissance Revivals

Medieval natural philosophy had been based largely on Aristotle, as mediated by Arabic commentators. During the fifteenth and sixteenth centuries, however, a period in retrospect loosely labeled as the Renaissance, a far wider range of classical writings was located, edited, translated if necessary, and gradually assimilated. The alternative currents of thought found in these writings played an important role in the development of the new natural philosophies of the seventeenth century. Two traditions were to be especially influential, that stemming from Aristotle's teacher Plato (c. 428–c. 348 B.C.), and the even older tradition of philosophical atomism. Both contained distinctive ideas about the nature of heat.

In contrast to Aristotle's mainly verbal, qualitative approach, Plato's natural philosophy emphasized the importance of mathematics, especially geometry. An inscription above the entrance to Plato's school, the Academy, is reputed to have read: "Let no one enter who does not know geometry." Although Plato accepted that the four elements were important components of the terrestrial world, he sought to provide a geometrical explanation for their existence and properties. He knew that there were five (and only five) regular or "Platonic" solids, which could conveniently be identified with the four terrestrial elements and the fifth celestial quintessence. Fire, for example, was composed of multitudes of minute tetrahedrons, because these had the smallest and sharpest (and thus most mobile and penetrating) shape.

Less influential in classical and renaissance times was atomism, the doctrine that the material world was composed exclusively of small, rigid, absolutely indivisible particles ("atoms") constantly and rapidly moving in an infinite, empty space or void. It was usually supposed that there were various kinds of atom, differing in shape and size. In the early fifteenth century, the long and detailed philosophical poem expounding atomist ideas, *On the Nature of Things,* by the Roman Lucretius (c. 99–c. 55 B.C.), was rediscovered.

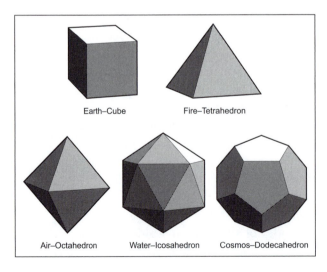

Earth–Cube   Fire–Tetrahedron

Air–Octahedron   Water–Icosahedron   Cosmos–Dodecahedron

Figure 1.2: The five possible regular or "Platonic" solids are the tetrahedron, cube, octahedron, icosahedron, and dodecahedron. Plato identified the first four with the four earthly elements, namely, fire, earth, air, and water. The fifth, the twelve-faced dodecahedron, formed from pentagons, he identified with a fifth element or "quintessence," of which he believed the Heavens were made. Illustration by Jeff Dixon.

The immediate impact of atomism was muted by its atheistic and hedonistic associations, but both atomism and Platonism were very influential in the early seventeenth century on the development of atomic and mechanical ideas in general and notions of heat in particular.

NEW EXPERIMENTS PHYSICO-MECHANICAL: HEAT IN THE SEVENTEENTH CENTURY

Heat is a very brisk agitation of the insensible parts of the object, which produces in us that sensation from whence we denominate the object hot; so what in our sensation is *heat*, in the object is nothing but *motion*. (English philosopher John Locke (1632–1704), quoted by Joule, 1850)

During the Renaissance European scholars were largely preoccupied with the recovery and assimilation of their complete classical heritage, insofar as they could find it. Only from the late sixteenth century were radical new philosophies of nature deliberately and persuasively developed and promoted, not only by Bacon, but most influentially by the Italian scientist Galileo Galilei (1564–1642), famous for his disagreements in later life with the Roman Catholic Church, and by the French philosopher Rene Descartes (or, des Cartes, whence "Cartesian") (1596–1650). The style of natural explanation that came to dominate the seventeenth century was atomistic or "mechanical," like the mechanism of a clock. As far as heat was concerned, this often resulted in assertions that "motion is the cause of heat," although these claims were generally very remote from the "kinetic" theory of heat that came to be widely accepted in the later nineteenth century (see chapter 6). Indeed, simultaneous frequent references to subtle fire-particles and aethers lent equal authority to the material, fluid theories that became increasingly popular through the eighteenth century. In retrospect, a more durable contribution was made by two specific strands of instrumental-experimental activity, firstly, the development of the thermometer and, secondly, the investigation of the properties of air using the vacuum pump.

New Natural Philosophies

Despite big differences in their recommended methods, there was considerable similarity between the theories of heat advanced by the new philosophers. Within the atomist-mechanical framework there were two main options for explaining heat, a material option and a dynamic option. The material option attributed the phenomena of heat to some all-pervasive *substance*, often similar or identical to the traditional element of "fire": The denser the substance in a given body, the greater the heat. The dynamic option regarded heat as the result of the rapid *motions* of the particles of ordinary matter: The more vigorous the motions, the greater the heat. More often than not, however, in the seventeenth century these two approaches were combined in various ways. Thus the phenomena of heat (and maybe also light) were widely attributed to the existence of particles of "fire," particles often so finely divided and tenuous as to constitute a fluid; but the thermal action of these fire particles was usually attributed to their rapidity of motion, either directly or as a result of the agitation that they imparted to grosser matter. Bacon, on the basis of his own "experimental history" of heat, concluded cautiously that "heat is a motion, expansive, restrained, and acting in its strife upon the smaller particles of bodies" (Bacon, 1620, bk. ii, aphorism xx, etc.). Galileo argued along Platonic lines that "those materials which produce heat in us and make us feel warmth, which are known by the general name of 'fire,' would then be a multitude of minute particles having certain shapes and moving with certain velocities. Meeting with our bodies, they penetrate by means of their extreme subtlety... The operation of fire by means of its particles is merely that in moving it penetrates all bodies, causing their speedy or slow dissolution in proportion to the number and velocity of the fire-corpuscles and the density or tenuity of the bodies" (Galilei, 1957 [1623], p. 277). Descartes' enormously influential system of natural philosophy maintained that matter (*res extensa* or "extended stuff") was identical with space itself, and thus essentially uniform and undifferentiated. To account for the undeniable variety of the experienced world, he argued that this uniform matter was capable of fragmentation into pieces of different shapes and sizes. There were a number of possible grades of fineness, the largest grade, similar in function to Lucretius' atoms, made up ordinary gross material objects. The finest grades of matter, indefinitely small "fire particles," formed the sun and the stars and, diffused throughout the world in the spaces between the larger particles, were responsible for the phenomena of heat in general. The wealthy Anglo-Irish nobleman and devout Puritan Robert Boyle (1627–91), often labeled "the father of modern chemistry," was a major exponent of the cautious, antispeculative, empirical methods promoted by Bacon. His *New experiments and observations touching cold, or an experimental history of cold, begun* was a model of the Baconian accumulation of data, the "natural history" of a topic. Even so, Boyle felt able to conclude that "...heat seems principally to consist in that mechanical property of matter we call motion" (quoted in Cardwell, 1971, p. 4), and went so far as to suggest that a rapid and random motion of constituent atoms was involved. The great English mathematician and scientist Sir Isaac Newton (1642–1727), without attempting a systematic treatment of the subject, similarly decided that heat was caused by the rapid motion of the parts of bodies.

Among the thermal phenomena discussed was the generation of heat by friction or hammering, often regarded in retrospect as crucial evidence for a mechanical theory of heat. But even if, like Galileo, one believed that motion (but specifically of fire particles) was the cause of heat, one might still conclude that "the rubbing together and friction of two hard bodies, either by resolving their parts into very subtle flying particles or by opening an exit for the tiny fire-corpuscles within, ultimately sets these in motion" (Galilei, 1957 [1623], p. 278), thus *releasing* (rather than generating) heat.

These materialist and dynamical theories of heat effectively eliminated "heat" as a real component of the material world. For Aristotle the qualities "hot" (and "cold") had been fundamental, more fundamental even than the elements such as "fire," which they served to define. Now instead the sensation of heat was a consequence of the shape, size, and speed of minute particles; there was no more "hot" (at least as we experience it) in a fire than there was "tickle" in a feather. This distinction between allegedly real "primary" qualities such as shape and size and merely mental "secondary" qualities such as taste and color came to be widely accepted. At the same time, the dichotomy between the supposedly distinct qualities of "hot" and "cold" was undermined. If heat were due to motion, there could be no *opposite* to heat: Cold—presumably an indication of sluggish motion—was simply the absence of heat. But Boyle, for one, was not entirely convinced; if there could be "fire" particles that melted things, then why not "cold particles" that caused their solidification?

Degrees of Heat and Cold: The Development of Thermometry

Seventeenth-century enthusiasm for "philosophical" instruments and experiments, combined with contemporary developments in the techniques of glass-working, resulted in the development of the "thermometer"—the name was first used in 1626. As already noted, medieval physicians were accustomed to think in terms of degrees of heat and cold as a measure of the "temperaments" of their patients, relying on their own senses as a guide. The first thermometers, providing a more objective measurement of "temperature," appeared around 1610 and were steadily refined over the following two centuries. The precise origins of these early thermometers are obscure. The most convincing claimant to their invention is probably the Italian physician Santorio Santorre (1561–1636), a friend and colleague of Galileo. In 1612 Santorre described a crude air thermometer, a glass bulb with a thin glass tube attached, filled with air but with the open end of the tube dipping into a vessel of water; the water would fall or rise in the tube as the air in the bulb expanded or contracted in response to changes in temperature of the surrounding air or of a patient's body. By the mid-1620s such instruments were already well-known and were often fitted with a numerical scale measuring degrees of heat and cold.

In the 1640s, it came to be recognized that these air thermometers would be affected by changes in atmospheric *pressure*, irrespective of any changes in temperature. There was a consequent shift to liquid-in-glass thermometers, in which the air in the bulb was replaced by a liquid, usually "spirits of wine" or alcohol; although the changes in volume were much smaller and more difficult to measure, they were unaffected by changes in pressure. From this it was only a small step to *sealed* liquid-in-glass thermometers identical in principle to modern thermometers. This step seems to have been taken before 1654 by Ferdinand II, Grand Duke of Tuscany, patron of the Accademia del Cimento

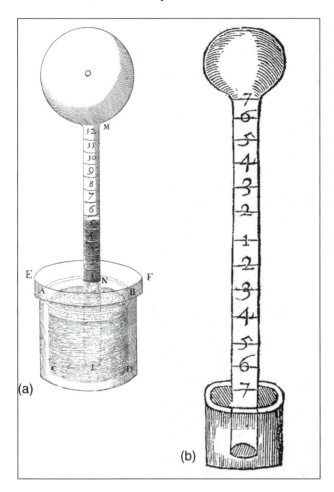

Figure 1.3: (a) Probably the earliest printed image of a thermometer with a numbered scale (of 12 degrees), in Robert Fludd's *Meteorologia Cosmica* (Frankfurt, 1626); (b) A slightly later image in his *Mosaicall Philosophy* (London, 1659 [1638]), this scale is graduated (following medical tradition) in 8 degrees each of hot and cold from a "temperate" midpoint. Both images: Wellcome Library, London.

(Academy of Experiments), the scientific society that flourished in Florence for a decade or so from the mid-1650s.

The early thermometers were provided with more-or-less arbitrary scales, often echoing the four or eight degrees of heat and of cold established in medical tradition. It was soon realized, however, that to achieve national and international consistency some objective, standardized scale was required. According to the Dutch scientist Christian Huyghens (1629–95), "It would be a good thing to think about a certain and universal measure of cold and heat; first make the capacity of the bulb have a certain proportion to that of the tube, and taking as a starting point the degree of cold at which water begins to freeze, or else the degree of heat of boiling water, so that without sending any thermometers, the degrees of heat and cold found in experiments can be communicated and consigned to posterity" (quoted in Middleton, 1966, pp. 50–51). Various schemes were proposed, all inevitably involving the choice of one or more "fixed points," that is, universally constant and reproducible temperatures. The use of snow (or freezing water) and boiling water—as in modern temperature scales—was suggested from the 1660s onwards, but in lively competition with numerous

Figure 1.4: The first sealed liquid-in-glass thermometers (Figs. I, II and IV), invented before 1654 by Ferdinand II, grand duke of Tuscany, patron of the Accademia del Cimento (Academy of Experiments) in Florence. Some of the credit should probably be given to the grand duke's extraordinarily skillful glass-worker, Mariani. Accademia del Cimento, *Saggi di naturali esperienze* [Essays of natural experiments] (Florence, 1666). Whipple Library, University of Cambridge.

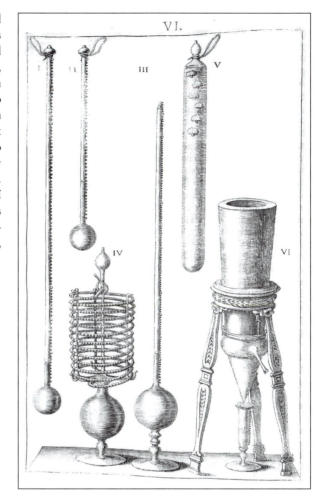

alternatives: The cellars of the Paris Observatory (or, indeed, any deep cellar) were widely regarded as being constant in temperature, "blood" (i.e., body) heat was widely adopted, as was the extreme cold that could be generated with a "freezing mixture" of salt and ice. But even when a pair of fixed points had been chosen, the numerical division of the intervening scale was still debatable. As a result, by the mid-eighteenth century there were at least a dozen competing scales in use.

Fortunately, only three of these survived much beyond the end of the century: those commonly known as the Fahrenheit, the Réaumur, and the Celsius (or centigrade) scales. The scale due to the German scientific-instrument maker Daniel Gabriel Fahrenheit (1686–1736) originally assigned zero degrees to the temperature of a mixture of ice, water and "sal-ammoniac [ammonium chloride] or also sea-salt," 32°F to a mixture of ice and water, and 96°F to the temperature reached "when the thermometer is held in the mouth, or under the armpit, of a living man in good health, for long enough to acquire perfectly the heat of the body..." (Fahrenheit, 1724, pp. 78–79). On such a scale the boiling point of water was *measured* as 212°F, although by 1740 it had become common practice to take this instead as the second fixed point. This "Fahrenheit" scale of temperature came into widespread use in England and Holland, partly because of the superior quality of Fahrenheit's thermometers. In France, on the other hand, a scale based on one devised around 1730 by the prestigious Academician Réne-Antoine de Réaumur (1683–1757) was widely adopted; this ran from 0°R at the freezing point to 80°R at the boiling point of water. In Sweden meanwhile in the early 1740s the astronomer Anders Celsius (1701–1744) suggested a scale of 100° between these same points, although initially his scale ran *downwards* from 0° at the temperature of boiling water to 100° in ice. Within a couple of years, however, this had been inverted to produce the familiar Celsius or centigrade scale, although this only came into widespread use, even in scientific circles, in the nineteenth century.

Newton made one minor contribution to thermometry. Available liquid thermometers could only measure relatively low temperatures, certainly not above the boiling point of mercury. In 1701 Newton argued that the *rate* of cooling of a given body would depend largely upon how hot it was: The hotter it was, the faster it would cool. Using his calculus Newton was able to express this relationship in precise mathematical terms. When an initially very hot body had cooled down to measurable temperatures, therefore, its rate of cooling could be measured, and thus, extrapolating back in time, its starting temperature could be estimated. Although mathematically elegant, Newton's "law of cooling" turned out not to be very accurate.

The Spring of the Air: Early Investigations of the Properties of Gases

In the nineteenth century, study of the physical properties of air and other gases played a crucial role in the development of ideas about heat (see chapters 2–6). That study itself began to take its modern shape in the seventeenth century. In the second quarter of the seventeenth century it was discovered that our atmosphere is effectively a vast, deep ocean of air and could therefore be analyzed using the concepts of pressure, buoyancy, and so on already developed in hydrostatics. The barometer, invented by Galileo's disciple Evangelista Torricelli (1608–47), could be used to measure the pressure of the atmospheric ocean; this turned out to be about the same as the pressure that would be exerted by a column of mercury some 30 inches high, although the precise figure was soon seen to vary depending on the weather and the altitude of the observer.

In order to investigate further, Boyle, with the help of such talented assistants as Robert Hooke (1635–1703), developed "a new pneumatical engine"—that is to say, an air (or "vacuum") pump connected to a glass bell-jar or "receiver," in which experiments could be performed; this apparatus became an icon of the new experimental science. There was one major physical difference between the oceans of the air and of the sea (other than their densities): A given amount of water had a more-or-less fixed volume, but a given body of air could be compressed into a smaller volume by a pump or, given the opportunity, would expand into a larger volume, just as a coiled spring could be squashed or stretched. Boyle's extensive studies of this "spring" and other properties of the air were published in 1660 in his *New Experiments Physico-Mechanical touching the spring of the air, and its effects; made, for the most part in a new pneumatical engine*. Although this work contains the relevant measurements on the relationship between the pressure and the volume of a body of air, it was only after his attention had been drawn to the earlier work of Henry Power (1623–68) that Boyle recognized the precise mathematical relationship involved, namely, "the pressures and expansions to be in reciprocal proportion" (Boyle, 1662, p. 60). So Boyle's law should probably be known as "Power's law"; on the continent it is often known as Mariotte's law, after the French scientist Edme Mariotte (c. 1620–84), who did not publish his findings until 1676, however.

Boyle's studies were conducted at constant temperature. The early air-thermometers made it evident, however, that air would expand or contract as its temperature increased and decreased. This phenomenon was first studied systematically by the Frenchman Guillaume Amontons (1663–1705), a member of the French Academy. He measured both changes in pressure at constant volume and changes in volume at constant pressure; in 1699 he recorded in both cases a quite accurate figure of increase by one-third between the temperatures of melting ice and boiling water.

Various explanations of the spring of the air were offered. Boyle himself speculated that the "Aerial Corpuscles" might be similar to the "slender and flexible hairs" in a fleece of wool, "each of which may indeed, like a little spring, be easily bent or rouled up; but will also, like a Spring, be still endeavouring to stretch itself out again" (Boyle, 1660, p. 22). Newton provided a much more precise and influential explanation. Proposition XXIII in Book II of the *Principia* gives a succinct and rigorous geometrical proof that, if air consisted of particles that repelled each other with "centrifugal forces…inversely proportional to the distances of their centers," then air would indeed "compose an elastic fluid" that would obey Boyle's law. Initially Newton was cautious to claim only that this was just an hypothesis; its mathematical elegance, however, combined with Newton's prestige, meant that the theory that air was composed of stationary, mutually repulsive particles became very widely accepted throughout the eighteenth century and beyond, although only with the rise of the caloric theory towards the end of the century was a plausible explanation of the repulsive force advanced.

A VERY "SUBTLE FLUID": HEAT IN THE EIGHTEENTH CENTURY

Whatever may be the cause that produces the sensation of heat, it is able to increase and decrease; and, from this point of view, it can be treated quantitatively. It does not seem that it occurred to the ancients to measure these relative values, and only in the past century have men conceived of ways of doing so. (Lavoisier, 1982 [1783], p. 3)

With a few notable exceptions, theories of heat during the eighteenth century increasingly turned away from the "dynamic" interpretations associated with Bacon, Boyle, and Newton and towards more "material" models that envisaged heat as a "subtle fluid." At the same time the early practical, commercial development of steam power began to overlap and (tentatively) interact with quantitative studies of specific and latent heats. In the last quarter of the century these materialist and quantitative trends, along with new ideas about the nature of gases, coalesced into the definitive "caloric" theory of heat that dominated the thinking of the following generation (see chapter 2).

Heat and Fire in the Mid-Eighteenth Century

One of the most influential natural philosophers of the early eighteenth century, through his teaching and his textbooks, was Hermann Boerhaave (1668–1738), professor of medicine, chemistry, and botany at the University of Leiden in Holland. His talent and influence lay in the pragmatic clarification and organization of ideas rather than in any startlingly original theories. He adopted an essentially Cartesian concept of heat as a "fire" material, composed of fine particles yet weightless; this material was distributed throughout space, and— and this probably reflects Boerhaave's more chemical interests—conceived as a universal agent of change. Boerhaave's notion of heat was therefore fundamentally materialist, yet *motion* continued to play an important role: The phenomena of heat were attributed to the vibrations of gross material particles, which vibration was in turn caused by the actions of the fire material. Either way, he made the important claim that his fire material was *conserved;* this provided a crucial foundation for further study of such issues as specific heat in the second half of the century.

From the 1740s onwards the concept of the "subtle fluid" became a popular— and fruitful—explanatory tool in many areas of physical science. This was especially evident in electricity where, from the mid-century onwards, speculative mechanical explanations were abandoned in favor of more descriptive accounts employing one or two electric fluids. These fluids—material and elastic yet "imponderable" (weightless)—were supposed to be self-repulsive but nevertheless strongly attracted to ordinary gross matter. In essence they were owed a great deal to contemporary concepts of "air"; the self-repulsive model of an elastic fluid developed by Newton was very widely accepted; recent developments in "pneumatic" chemistry had introduced the additional notion that apparently enormous volumes of "air" could be "fixed" in solids (such as carbonates), suggesting that air had a capacity to attach itself to ordinary matter. It is not surprising that by the late 1770s a number of writers had outlined theories of heat or "fire" as a subtle, imponderable, elastic fluid; elastic because self-repulsive, but also attracted to ordinary matter. The definitive "caloric" version of such a theory appeared in the early 1780s.

The main exception to the materialist trend was the Swiss mathematician Daniel Bernoulli (1700–82). In his *Hydrodynamics* of 1738 he developed a theory of elastic fluids (such as air) that was similar in its basic assumptions to the "kinetic theory of gases" that came to be generally accepted in the later nineteenth century (see chapter 6). Bernoulli assumed that air was composed of "very small particles in very rapid motion" (Bernoulli, 1968 [1738], p. 200), exerting pressure through their collective impacts on the walls of any container. On this basis Bernoulli was able to give a mathematical derivation of Boyle's law.

He further related the increase of air pressure as temperature increased to "more intense motion of its particles...." (ibid., p. 202), deducing more precisely that the pressure of air "will vary as the square of the particle velocity...." Even so, Bernoulli does not claim that temperature is directly connected to molecular velocities or kinetic energies. Rather he suggests that the definition of temperature scales "is arbitrary and not inherent in nature...." Nevertheless, he says, "it seems reasonable to me to determine the temperature of air by reference to its pressure, provided it is of standard density" (ibid., 204). Over the next century there was little response to Bernoulli's speculations. At the time his approach lacked any clear advantages over Newton's well-established and authoritative repulsion theory of elastic fluids, and it was not at all clear how Bernoulli's molecular motions related to the phenomena of heat in general. In particular, it did not offer a conservative foundation for quantitative investigation as provided by Boerhaave and the subtle fluid theories.

Heat and the "Fire Engine"

I believe that, if all the vis viva which is hidden in a cubic foot of coal and is drawn out of it by burning were usefully expended in operating machines, more could be achieved thus than by a day's labour by 8 or 10 men. (Daniel Bernoulli, 1968 [1738], p. 231)

The development of thermodynamics in the nineteenth century was very closely connected with contemporary efforts to understand the operation of steam engines, which were crucial to the industry and transport of the age. The serious industrial application of steam power began much earlier, however, when in 1698 Thomas Savery (1650?–1715) was granted a patent for "Raiseing of water...by the Impellent Force of Fire." In 1702 Savery described his engine in a booklet entitled "The Miners Friend; or, an Engine to raise Water by Fire, described. And of the manner of Fixing it in Mines...." A much more effective "fire engine" was designed by the West Country ironmonger Thomas Newcomen (1663–1729) in the first decade of the eighteenth century. Both the Savery and Newcomen engines were "atmospheric" engines, depending upon the pressure of the atmosphere to function. Numerous Newcomen engines were erected to pump water out of mines throughout Britain in the eighteenth century.

A major improvement in the design of the steam engine was the introduction of the separate "condenser," patented in 1769 by James Watt (1736–1819), at that time working as instrument maker to Glasgow University. In a Newcomen engine, the steam was condensed by a jet of cold water inside the main cylinder. Watt saw that the consequent repeated heating and cooling of the cylinder was wasteful of heat. He realized that the condensing of the steam could equally well take place in a separate adjacent chamber or "condenser": The cylinder could thus remain permanently hot and the condenser permanently cool, with considerable economy of steam and fuel. And there was a further refinement: In the Newcomen engine it was cold, atmospheric *air* that drove the piston down on the power stroke, once again contributing to an unwelcome cooling of the cylinder; Watt realized that this cold air could be replaced by (hot) steam, which would not cool down the cylinder. Thus the "atmospheric" engine became properly a "steam" engine for the first time.

The fundamental incentive to this development of steam power was the expansion of the British economy from the end of the seventeenth century onwards;

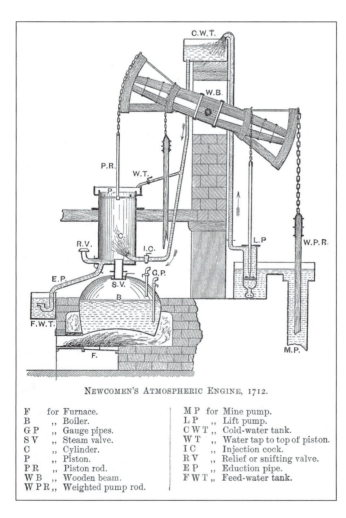

NEWCOMEN'S ATMOSPHERIC ENGINE, 1712.

F	for	Furnace.	M P	for	Mine pump.
B	,,	Boiler.	L P	,,	Lift pump.
G P	,,	Gauge pipes.	C W T	,,	Cold-water tank.
S V	,,	Steam valve.	W T	,,	Water tap to top of piston.
C	,,	Cylinder.	I C	,,	Injection cock.
P	,,	Piston.	R V	,,	Relief or snifting valve.
P R	,,	Piston rod.	E P	,,	Eduction pipe.
W B	,,	Wooden beam.	F W T	,,	Feed-water tank.
W P R	,,	Weighted pump rod.			

Figure 1.5: Thomas Newcomen's atmospheric "fire" engine of about 1712, the first commercially successful steam pump. Image from Andrew Jamieson, *Elementary Manual on Steam and the Steam Engine* (London, 1900), by permission of the Syndics of Cambridge University Library.

Figure 1.6: James Watt's more efficient engine of 1769, incorporating a separate condenser. Image from Andrew Jamieson, *Elementary Manual on Steam and the Steam Engine* (London, 1900), by permission of the Syndics of Cambridge University Library.

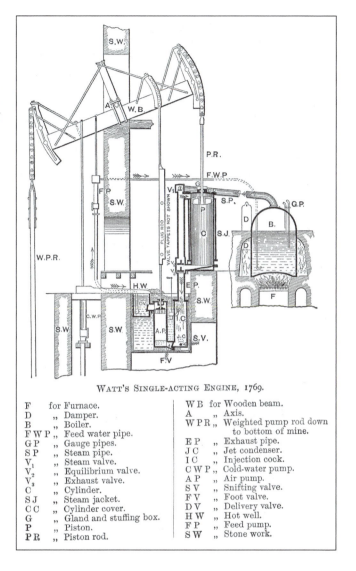

WATT'S SINGLE-ACTING ENGINE, 1769.

F	for Furnace.		W B	for Wooden beam.
D	,, Damper.		A	,, Axis.
B	,, Boiler.		W P R	,, Weighted pump rod down
F W P	,, Feed water pipe.			to bottom of mine.
G P	,, Gauge pipes.		E P	,, Exhaust pipe.
S P	,, Steam pipe.		J C	,, Jet condenser.
V_1	,, Steam valve.		I C	,, Injection cock.
V_2	,, Equilibrium valve.		C W P	,, Cold-water pump.
V_3	,, Exhaust valve.		A P	,, Air pump.
C	,, Cylinder.		S V	,, Snifting valve.
S J	,, Steam jacket.		F V	,, Foot valve.
C C	,, Cylinder cover.		D V	,, Delivery valve.
G	,, Gland and stuffing box.		H W	,, Hot well.
P	,, Piston.		F P	,, Feed pump.
P R	,, Piston rod.		S W	,, Stone work.

increased demand for tin and coal created a market for improved methods of draining mines—hence Savery's *The Miners Friend*. The extent of the linkage between practical, technological innovation and contemporary developments in science is far from clear, however. In the tradition established by Bacon, the Royal Society was certainly eager to emphasize the potential *usefulness* of the new experimental philosophy, and Savery demonstrated his engine to them in 1699. The fact that Amontons' paper in the same year was entitled "[A] Method of Substituting the Force of Fire for Horse and Man Power to Move Machines" suggests that the French Academy of Sciences had similar interests. On the other hand, Newcomen seems to have been a relatively unlearned artisan who developed an improved engine by trial and error, largely unaided by the speculations of more learned scientists in London and Paris.

Either way, keen interest in the steam engine remained a feature of both academic and popular Newtonian natural philosophy in the first half of the eighteenth century. Thus it was that in 1763 Watt was asked to repair the model atmospheric engine used in the natural philosophy lectures at Glasgow University, a task that resulted in the improvements and patents noted above and his subsequent shift into industry. The debt that Watt might have owed to the contemporary speculations of Black on specific and latent heats remains debatable, however. There is no doubt that the two men, both working in the 1760s at Glasgow University, were well acquainted and discussed matters of mutual interest. Watt, however, was adamant that his technical improvements owed little if anything to Black's more theoretical inquiries.

Specific and Latent Heats

> The quicksilver, therefore, may be said to have less capacity for the matter of heat. And we are thus taught, that, in cases in which we may have occasion to investigate the capacity of different bodies for heat, we can learn it only by making experiments. Some have accordingly been made, both by myself and others. (Black, 1803 [c.1760], vol.1, p. 82; quoted in McKie and Heathcote, 1935, p. 15)

Although the idea that heat was an indestructible (and therefore conserved) material was increasingly common in the mid-eighteenth century, questions about the variable distribution of this material had barely been articulated, much less convincingly answered. The concept of "quantity of heat"—as distinct from the intensity of heat, or temperature—was rarely discussed. Late seventeenth- and early eighteenth-century attempts to construct less-arbitrary temperature scales sometimes involved the mixture of quantities of water at different temperatures. The English natural philosopher and mathematician Brook Taylor (1685–1731), for example, mixed equal amounts of boiling and freezing water to verify the mid-point of his thermometer's scale. Such procedures seem to involve an implicit concept of a quantity of heat proportional to the temperature and also to the weight or volume of a body. As already noted, Boerhaave finally helped to clarify the problem, if not its solution, by explicit insistence that the quantity of heat (or matter of fire) must be conserved. More originally, he also reported experiments, probably conducted for him by Fahrenheit, on the mixture of volumes of *different* substances, namely water and mercury, at different temperatures; the results seemed to suggest that, far from being proportional to the weight of a body, the amount of heat in it was more nearly proportional to its *volume*.

It was Joseph Black (1728–99) who first appreciated, sometime around 1760, that different materials might each have a characteristic "capacity" for heat—a "specific heat," as it was soon to be called, different from either their volume or their weight. Black was professor of medicine and chemistry at Glasgow and then at Edinburgh; he was a first-rate scientist and a brilliant lecturer, very influential on his many pupils and colleagues, but he unfortunately neglected to publish much after his doctoral dissertation; his lectures were only officially published posthumously, edited by his former pupil and colleague

John Robison (1739–1805). Reflecting on the experiments reported by Boer-haave, Black concluded that they showed that the mercury "has less *capacity* for the matter of heat than water (if I may be allowed to use this expression) has; it requires a smaller quantity of it [heat] to raise its [mercury's] temperature by the same number of degrees" (Black, 1803, p. 81; quoted in McKie and Heath-cote, 1935, p. 14). This capacity could only be determined by experiment (see quote at head of section). According to an editorial note by Robison:

> Dr. Black estimated the capacities, by mixing the two bodies in equal masses, but of different temperatures; and then stated their capacities as inversely pro-portional to the changes of temperature of each by the mixture. Thus, a pound of gold, of the temperature 150°[F], being suddenly mixed with a pound of water, of the temperature 50°, raises it to 55° nearly: Therefore the capacity of gold is to that of an equal weight of water as 5 to 95, or as 1 to 19; for the gold loses 95°, and the water gains 5°. (Black, 1803, p. 506; quoted in McKie and Heathcote, 1935, p. 15)

This suggests a relative specific heat of 0.053 (the modern figure is 0.031). But Black himself gave no further details and undertook no systematic inves-tigations. A table of some 40 specific heats, mainly due to the chemist and geologist Richard Kirwan (1733–1812), was published in 1780. A much more detailed and systematic treatment of the whole subject was presented in 1781 by the Swedish scientist Johan Carl Wilcke (1732–96).

Black (from 1757) and Wilcke (in 1772) also recognized, probably inde-pendently, the existence of a hidden or "latent" heat involved in changes of state, such as melting or boiling. Melting ice and just melted "ice-cold" water, for example, have the same temperature. Previously, therefore, it had been as-sumed that, when a substance reached its melting or boiling point, only a very small amount of extra heat was required to complete the fusion or vaporization. Reflection upon such phenomena as the quite gradual thawing of snow and ice in the springtime led both Black and Wilcke to realize that this assumption must be wrong. Black conducted a series of experiments on the *time* taken to thaw a given mass of ice, which he compared with the rise in temperature of an equal mass of ice-cold water over the same period. He concluded that sub-stantial amounts of heat were indeed required to bring about both changes and gave reasonable estimates of the quantities involved. Wilcke, using a method of mixtures, arrived at a similar result for the melting of ice. In the early 1760s, in the course of his more practical investigations into the steam engine, Watt also noticed the excess heat contained in steam and measured it by bubbling steam through cold water. "In both cases," commented Black, "considered as the cause of warmth, we do not perceive its [heat's] presence: it is concealed, or latent, and I gave it the name of LATENT HEAT" (Black, 1803, p. 157; quoted in McKie and Heathcote, 1935, p. 22).

There remained the question of what had happened to the heat absorbed in both cases, heat that had apparently disappeared without causing any perceptible or "sensible" change in temperature. An ingenious and influential solution connecting latent and specific heats was formulated by Black's assistant William Irvine (1710–90) and developed by Irish surgeon Adair Crawford (1748–95). Irvine maintained that there was no such thing as hidden or "combined" heat, only free heat: The heat emitted or absorbed upon a change of state was the result of a sudden change in the specific total heat capacity of the material in question. (Since water, for example, had a greater specific heat than ice, as experiment confirmed, then on melting extra heat would be needed to supply the sudden increase in total heat capacity. The logic of Irvine's argument required the existence of an absolute zero of temperature corresponding to a total absence of heat.) The heat released in chemical reactions was similarly attributed to changes in the specific heats of the reagents. Crawford's 1779 *Experiments and observations on animal heat,* which contained extensive if rough experimental data to back the case, was very influential for several decades; widely accepted in Britain, in France it was often seen as a rival to the theories of Lavoisier. Either way, by about 1780, the concept of the quantity of heat in a body had become well established, which helped to consolidate commitment to the material view of heat.

Lavoisier and the Crystallization of Caloric

The name of Antoine Laurent Lavoisier (1743–94) is most commonly associated with a revolution in chemistry, often identified with a new "oxygen" theory of combustion. Born into a wealthy family of lawyers, the able and ambitious young Lavoisier pursued a career in science from a young age. As early as 1768 he managed to be elected to fellowship of the French Academy; as a fellow of the Academy he assisted in the preparation of official reports on such varied topics as the water supply to Paris, the manufacture of gunpowder, and the Montgolfier balloon. In the same year he consolidated his financial interests by purchasing a share in the private company of "tax farmers" who, very lucratively, collected customs duty on tobacco, salt, and imported goods for the French government. Three years later Lavoisier furthered both his scientific and his financial careers by marriage to Marie-Anne Pierette Paulze (1753–1836), the daughter of a fellow tax-farmer. Although only fourteen at the time of their marriage, she assisted him in his scientific work, translating scientific papers from English, helping with experiments, and illustrating apparatus for his textbooks. Lavoisier's new oxygen theory of combustion was gradually worked out during the 1770s and definitively presented in his *Elementary Treatise of Chemistry* in 1789. Although he was vigorously involved in the reform programs of the early French Revolution, his association with the extremely unpopular tax-farmers sadly resulted in Lavoisier's execution by guillotine in 1794.

But Lavoisier's new chemistry was intimately connected to new ideas about the nature of gases and heat in general. His approach to the study of heat was first explained in detail in a 1783 *Memoir on Heat,* written jointly with his younger, more mathematical colleague Pierre Simon de Laplace (1749–1827), but also figured extensively in the *Elementary Treatise*, the very first chapter of which is "On the Combinations of Caloric, and the Formation of Elastic Aëriform Fluids or Gasses." Lavoisier's exposition of the "caloric" theory provided a framework for the study of heat for the next generation, especially in France, providing an essentially material—indeed "chemical"—and quantitative concept of heat.

Basically, Lavoisier took the widely established but rather qualitative concept of heat as a "subtle fluid" and gave it extra coherence and precision, integrating it with the new concepts of specific and latent heat, and with new ideas about gases. The basic effects of heat, he noted, were to cause expansion and changes of state. "It is difficult to comprehend these phenomena, without admitting them as the effects of a real and material substance, or very subtle fluid, which, insinuating itself between the particles of bodies, separates them from each other" (Lavoisier, 1790, p. 4). Initially labeled "igneous fluid" or "matter of fire," by 1787 this substance had been named "caloric" by Lavoisier.

This caloric could exist in both "free" and "combined" states. Changes in the density of free caloric in a given body were perceptible ("sensible") and measurable as changes in temperature, although the precise change in temperature depended upon the specific heat of the material of the body. In the combined state, on the other hand, the presence of the caloric would result in no detectable change in temperature, but might instead be responsible for a change of state, from solid to liquid or from liquid to gas. According to Lavoisier, it was "right to assume, as a general principle, that almost every body in nature is susceptible of three states of existence, solid, liquid, and aëriform, and that these three states of existence depend upon the quantity of caloric combined with the body. Henceforwards I shall express these elastic aëriform fluids by the generic term gas" (ibid., 15). The quantity of caloric required to effect a given change of state was, of course, equivalent to the "latent heat" identified by Black and Wilcke.

But caloric could also be combined in various amounts with different materials and could be released (or, indeed, absorbed) in the course of chemical reactions. It was the caloric contained in oxygen gas that was released as heat and flame in the process of combustion; Lavoisier's new caloric thus took over many of the attributes and functions of the "phlogiston" previously regarded as the "principle of fire" contained in inflammable substances. Overall, Lavoisier's caloric was very much a *chemical* substance, weightless but combining with other grosser substances in a variety of ways. Indeed, "caloric" (along with "light") appears at the very top of Lavoisier's "Table of simple substances"; its "correspondent old names" are given as "Principle or element of heat. Fire. Igneous fluid. Matter of fire and of heat" (Lavoisier, 1790, p. 175).

In principle, especially in the earlier *Memoirs*, Lavoisier (and Laplace) claimed to be undecided about the ultimate nature of heat, whether material

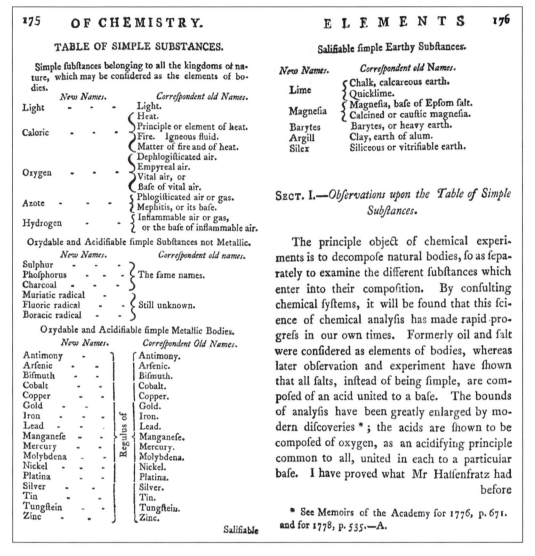

175   OF CHEMISTRY.   ELEMENTS   176

TABLE OF SIMPLE SUBSTANCES.

Simple fubftances belonging to all the kingdoms of nature, which may be confidered as the elements of bodies.

New Names.		Correfpondent old Names.
Light - - -		Light.
Caloric - - -		Heat. / Principle or element of heat. / Fire. Igneous fluid. / Matter of fire and of heat.
Oxygen - - -		Dephlogifticated air. / Empyreal air. / Vital air, or / Bafe of vital air.
Azote - - -		Phlogifticated air or gas. / Mephitis, or its bafe.
Hydrogen - - -		Inflammable air or gas, / or the bafe of inflammable air.

Oxydable and Acidifiable fimple Subftances not Metallic.

New Names.	Correfpondent old names.
Sulphur - -	
Phofphorus - - -	The fame names.
Charcoal - -	
Muriatic radical -	
Fluoric radical - -	Still unknown.
Boracic radical - -	

Oxydable and Acidifiable fimple Metallic Bodies.

New Names.	Correfpondent Old Names.
Antimony -	Antimony.
Arfenic - -	Arfenic.
Bifmuth - -	Bifmuth.
Cobalt - -	Cobalt.
Copper - -	Copper.
Gold - -	Gold.
Iron - - -	Iron.
Lead - -	Lead.
Manganefe - -	Manganefe.
Mercury - -	Mercury.
Molybdena - -	Molybdena.
Nickel - -	Nickel.
Platina - -	Platina.
Silver - -	Silver.
Tin - -	Tin.
Tungftein - -	Tungftein.
Zinc - -	Zinc.

Regulus of

Salifiable

Salifiable fimple Earthy Subftances.

New Names.	Correfpondent old Names.
Lime	Chalk, calcareous earth. / Quicklime.
Magnesia	Magnefia, bafe of Epfom falt. / Calcined or cauftic magnefia.
Barytes	Barytes, or heavy earth.
Argill	Clay, earth of alum.
Silex	Siliceous or vitrifiable earth.

SECT. I.—*Obfervations upon the Table of Simple Subftances.*

The principle object of chemical experiments is to decompofe natural bodies, fo as feparately to examine the different fubftances which enter into their compofition. By confulting chemical fyftems, it will be found that this fcience of chemical analyfis has made rapid progrefs in our own times. Formerly oil and falt were confidered as elements of bodies, whereas later obfervation and experiment have fhown that all falts, inftead of being fimple, are compofed of an acid united to a bafe. The bounds of analyfis have been greatly enlarged by modern difcoveries * ; the acids are fhown to be compofed of oxygen, as an acidifying principle common to all, united in each to a particular bafe. I have proved what Mr Hatfenfratz had before

* See Memoirs of the Academy for 1776, p. 671. and for 1778, p. 535.—A.

Figure 1.7: Antoine Lavoisier's table of "simple substances" or chemical elements, which starts with light and caloric (heat); from Lavoisier, *Elements of Chemistry* (Edinburgh, 1790), translation of his *Traité élémentaire de chimie* (Paris, 1789), by permission of the Syndics of Cambridge University Library.

or dynamic. They insisted, however, that in *either* case, "If, in any combination or change of state, there is a decrease in free heat, this heat will reappear completely whenever the substances return to their original state; and conversely, if in the combination or in the change of state there is an increase in free heat, this new heat will disappear on the return of the substances to their original state" (Lavoisier, 1982 [1783], p. 6). In other words, the total quantity of heat, whether matter or motion, was always *conserved* in all processes—although, for future reference, it should be noted that this also required that heat could neither be created (for example, by friction) nor be destroyed (and converted into work, as in a steam engine).

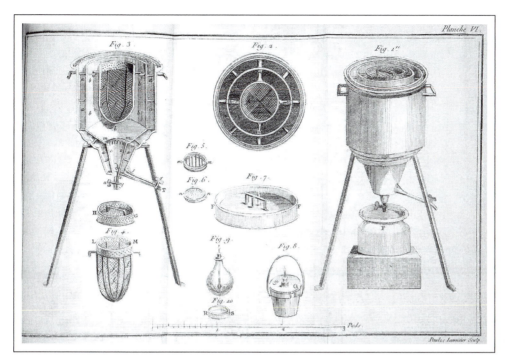

Figure 1.8: The ice-calorimeter invented by Antoine Lavoisier and Pierre-Simon Laplace in 1783 to enable precise quantitative measurement of specific heats, heats of chemical reaction, and so forth; from Lavoisier, *Traité élémentaire de chimie* (Paris, 1789), Whipple Library, University of Cambridge.

This principle of the conservation of heat provided a secure foundation for the quantitative study of heat, a study that was embodied in Lavoisier and Laplace's newly invented experimental apparatus, the ice "calorimeter." The method of mixtures used by Black and Wilcke was quantitative, but was limited in its scope; it could not easily be used with materials that did not mix or, worse, reacted chemically. Use of the ice-calorimeter, on the other hand, "extends to all phenomena in which there is the production or absorption of heat" (ibid., p. 12). The basic principle was to measure quantities of heat by weighing the amount of ice that they could melt. With this apparatus Lavoisier and Laplace measured a number of specific heats, but also the amount of ice melted (and thus the heat released) in a range of chemical reactions—for example, "By detonating 1 ounce of saltpeter with 1/3 ounce of carbon: Quantity of ice melted—12 ounces" (ibid., p. 18).

Among eighteenth-century scientists (or "physicians") interest in combustion was closely related to efforts to understand the processes of human and animal respiration—Crawford's influential work, for example, was specifically on "animal heat." It is not surprising, therefore, to find Lavoisier also pursuing this topic:

The room temperature being 1 1/2 degrees [Réaumur], we put in one of our calorimeters a guinea pig whose body temperature was about 32 degrees and consequently little different from that of the human body. So that it would not suffer during the experiment, we placed it in a little basket lined with cotton, whose temperature was zero. The animal remained in the apparatus 5 hours and 36 minutes.... The well drained apparatus furnished about 7 ounces of melted ice... The animal did not appear to suffer in these experiments. (ibid., p. 19)

Lavoisier's account of heat also presented a clear "caloric theory of gases." The Newtonian hypothesis of mutually repulsive gas particles had been widely accepted, the repulsion often referred to some kind of "atmosphere," maybe electrical, around each particle. The more recent recognition of gas as matter in a *gaseous state*, thanks to the influence of heat, suggested that an atmosphere of *heat* might be responsible instead. Thus, according to Lavoisier, "It is by no means difficult to perceive that this elasticity [of gasses] depends upon that of caloric, which seems to be the most eminently elastic body in nature" (Lavoisier, 1790, p. 22). Indeed, caloric itself, albeit weightless, must be an

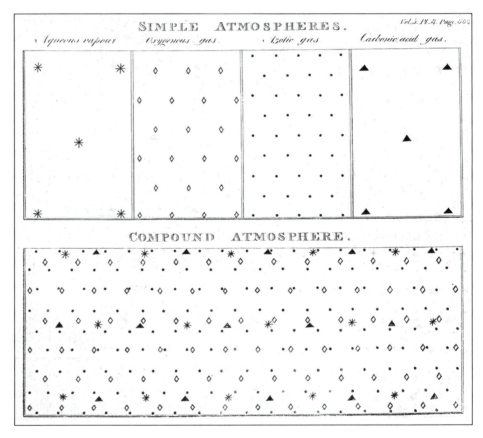

Figure 1.9: The widely accepted late eighteenth-century theory that gases (including mixed or "compound" gases) consist of static arrays of mutually repulsive molecules, as represented in John Dalton's "On the expansion of elastic fluids by heat," *Memoirs of the Literary and Philosophical Society of Manchester*, 5, part 2 (1802); by permission of the Syndics of Cambridge University Library.

elastic fluid just like a gas, and have the very same structure as a gas: "We have very good right to conclude, that the particles of caloric mutually repel each other" (ibid., p. 23), although they must also have a powerful attraction for particles of ordinary matter. Similar ideas had been advanced somewhat obscurely by other scientists in the late 1770s, but Lavoisier's clear and concise presentation of the theory in his revolutionary chemistry textbook was enormously more influential.

The year of publication of Lavoisier's definitive *Elements of Chemistry*, 1789, also saw the start of the French Revolution (1789–94), the socio-political paroxysm often taken to mark the origin of the modern centralized, secular state. As an excise tax-collector under the Old Regime Lavoisier had tried to impose strict controls on the flow of taxable goods into and out of Paris. As a scientist he had developed a coherent experimental and theoretical framework to impose a strict quantitative accounting upon the flow of heat in and out of bodies. (For most of his contemporaries, what eventually came to be seen as loopholes in his system—radiant heat, the generation of heat by friction, least of all the consumption of heat in a steam engine—only became problematic much later.) Over the ensuing quarter century of the Revolutionary (1789–94) and Napoleonic (1794–1815) periods, caloric theories provided a solid and fruitful foundation for further work, especially on the thermal properties of gases.

2

THE RISE AND FALL OF THE CALORIC THEORY

INTRODUCTION: THE GOLDEN AGE OF FRENCH SCIENCE

By means of these assumptions, the phenomena of expansion, heat and vibrational motion in gases are explained in terms of attractive and repulsive forces which act only over insensible distances.... All terrestrial phenomena depend on forces of this kind, just as celestial phenomena depend on universal gravitation. It seems to me that the study of these forces should now be the chief goal of mathematical philosophy. (Laplace, 1799–1825, vol. 5, p. 99; Fox, 1990, p. 285)

The High-Flying Scientific Career of Joseph Gay-Lussac

On September, 16, 1804, the young French scientist Joseph Louis Gay-Lussac (1778–1850) took off in a hydrogen balloon; starting in Paris he eventually landed near Rouen, some 70 miles (110 km) away, having reached a record height of 7,016 m (23,000 feet), a record that remained unbroken until the year of his death—and even then surpassed only by a few meters. The age of lighter-than-air flight had opened some 20 years earlier with the hot-air balloon flights of the Montgolfier brothers in 1783, closely followed later that year by the hydrogen balloon ascents of Jacques Alexandre César Charles (1746–1823). The motivation of the early balloonists was mainly adventure and spectacle. The flights undertaken by Gay-Lussac, on the other hand, were devoted to serious scientific study: He made systematic measurements of the Earth's magnetic field and of variations in the pressure, temperature, humidity, and composition of the atmosphere. Gay-Lussac was one of the first of a new breed of highly trained professional scientists produced by the revitalized scientific institutions of post-Revolutionary France. He had already published an important study of the thermal expansion of gases, leading to the unequivocal

LE BALLON DE LYON, *nommé le de Flesselles de 120 pieds de haut sur 102 de diametre, d'après les dimensions données par*
Mr. Montgolfier et élevé des Brotteaux jusqu'après de 1400 Toises, le 19 Janvier 1784. monté par
M.M.

Montgolfier l'ainé.
Pilastre du Rozier.
Le Prince Charles de Ligne.
Le Comte de la Porte d'Anglefort.

Elevés vous, audacieux mortels,
Dans la plaine des airs, sur l'aile de la gloire,
C'est à vous seuls d'aller au Temple de Mémoire
Ecrire vos noms immortels.

M.M.
Le Comte de Laurencin.
Le Comte de Dampiere.
et le Sr. Fontaine.

Figure 2.1: The first manned flights in the hot-air balloons of the Montgolfier brothers took place in Paris in the autumn of 1783. This image represents an ascent at Lyon in January 1784 in the then-largest balloon built, which succeeded in carrying seven occupants to a height of 3,000 feet. Library of Congress.

statement of a law that is often, rather unfairly, attributed to Charles—there was indeed a close connection between the study of gases and ballooning. Gay-Lussac was subsequently credited with a string of scientific discoveries and practical commercial processes. The eponymous "law of combining volumes," crucial to the subsequent development of the chemical atomic theory, claimed that the volumes of gases involved in chemical reaction, if measured at the same temperatures and pressures, are always in simple whole-number ratios. "Gay-Lussac towers" became an essential component of the efficient commercial production of that index of industrial development, sulfuric acid. Gay-Lussac was to become one of the leading lights of a "Golden Age" of French science. It was within this French context, in fact, that most of the major developments in the understanding of heat took place in the first quarter of the nineteenth century.

Revolution and the Reform of French Science

The French Revolution lasted from 1789 to 1794 and included the execution of King Louis XVI in 1793 and the "Terror" of 1793–4. The young general Napoleon Bonaparte (1769–1821) seized power in 1799, installing himself as emperor in 1804, which he remained until his defeat at the battle of Waterloo in 1815. Alongside the bloodshed and terror France was transformed from an absolute monarchy, supported by a privileged aristocratic-ecclesiastical elite, into a modern secular state run, in principle at least, on rational, central, and egalitarian lines. The successive revolutionary governments legislated a host of fundamental reforms to French customs and institutions. Typical, even symbolic, was the reform of the various traditional systems of weights and measures by the introduction of the single, rational, unified, centralized, standardized decimal "metric" system; the meter, gram, and liter are still with us, of course, although the system for time—with 10 days per week, 10 hours per day, and so on—proved less durable.

Scientific institutions were reformed along with everything else. The prestigious (but aristocratic) Academy of Sciences was reconstituted as part of a newly created cultural body, the Institute of France; the so-called "First Class of the Institute" was devoted to scientific research and debate. The École Polytechnique (Polytechnic School) was founded in 1794 to give an elite corps of scientists and engineers the best mathematical and experimental training available, including systematic work in teaching laboratories. Graduates of the École went on to further study in more specialized state schools of mines, of artillery, of roads and bridges, and so forth. Access to these new institutions, in principle at least, was now independent of wealth or breeding and was based on merit assessed by examination. Thus there emerged for the first time a coherent career structure for a scientist, from training at the École Polytechnique and other more specialized state technical schools, through teaching at these institutions, hopefully to be elected eventually to membership of the Institute. It was possible to be a *professional* "scientist," rather than a physician, a mathematician, a natural philosopher, a clergyman, or an independently wealthy gentleman with an interest in natural science.

The Laplace Program

In the first quarter of the nineteenth century the new Revolutionary institutions delivered a "Golden Age" for French science, associated with the naturalists Jean Lamarck (1744–1829) and Georges Cuvier (1769–1832) and a flotilla of physicists, aspects of whose work we shall study further in this chapter and the next.

In the Napoleonic era (1799–1815) the physical sciences were dominated by two scientists in particular, the astronomer-mathematician (and Lavoisier's collaborator in the study of heat) Laplace and the chemist Claude Louis Berthollet (1748–1822). Laplace and Berthollet had a very definite program for the physical sciences, which they vigorously promoted both in their own research and teaching and through their influence within the scientific establishment.

At the methodological level, Laplace and Berthollet were eager to encourage the integration of experimental and mathematical science, creating a discipline close to modern mathematical physics. Throughout most of the eighteenth century the mathematical sciences (mechanics, optics, astronomy) were only loosely engaged with experimental natural philosophy (electricity, chemistry, and so on). According to another ambitious young protégé of Laplace, Jean Baptiste Biot (1774–1862), writing in 1806, "People have been content to offer the public a certain series of brilliant experiments rather than try to fix exactly the laws of the phenomena and determine their relationships, which can only be done by mathematical reasonings" (Frankel, 1977, p. 45). Towards the end of the eighteenth century French scientists in particular, often with the benefit of sophisticated mathematical training in the state engineering schools, combined the use of custom-built precision experimental apparatus with sophisticated mathematical analysis of the measurements. The ice-calorimeter illustrated this trend, which Laplace and Berthollet were now to consolidate. According to Laplace, also in 1806:

> The need for very precise experiments increases as the sciences become more perfect....One cannot encourage physicists too strongly to give the greatest precision to their results; similarly, one cannot give enough encouragement to a skillful artisan dedicated to the perfection of scientific instruments. One badly done experiment has often been the cause of many errors; while a well-executed experiment lasts forever and often becomes the source of new discoveries. (Frankel, 1977, p. 44)

The result was an increasingly refined experimental culture of precision measurement that led the world for a generation.

Laplace and Berthollet also had a preferred style when it came to the interpretation and explanation of experimental results. In Newtonian astronomy, everything was explained in terms of gravitational forces acting between material bodies. In the same way, according to Laplace, when trying to understand heat or any other phenomena in the terrestrial domain, one should think in terms of very short-range attractive and repulsive forces acting between the microscopic particles of matter. Berthollet agreed: "The forces that bring about chemical phenomena all derive from the mutual attraction between the molecules of bodies. The name affinity has been given to this attraction so as to distinguish it from astronomical attraction.... It is probable that both are one and the same property" (Berthollet, 1803, i, p. 1; Fox, 1990, p. 281).

But Laplace and Berthollet's influence was not just intellectual. They wielded great power within the Parisian scientific establishment. The problems set for the Institute's regular prize competitions often reflected their interests, and the selection of the winner could be influenced by their preconceptions. Their patronage could be crucial to career advancement; ambitious protégés like Gay-Lussac and Biot would often undertake research that was focused on Laplace and Berthollet's interests and interpreted in the light of their paradigm. An inner circle of disciples came to meet regularly at Berthollet's home at Arcueil outside Paris, where there were extensive laboratory facilities; the semi-formal "Society of Arcueil" thus constituted an early elite research institute. More negatively, those whose methods and theories clashed with the Laplace program—men like Joseph Fourier (1768–1830) and Augustin Fresnel (1788–1827)—could find themselves marginalized, although this was less the case after the fall of Napoleon in 1815.

Laplace and Berthollet's program also included a number of more specific theoretical commitments. Laplace, for example, was committed to the corpuscular (as opposed to the wave) theory of light, and Berthollet rejected the constant composition of chemical compounds—and thus the chemical atomic theory when it was proposed.

Heat in the Early Nineteenth Century: Outline of Issues and Achievements

From the end of the eighteenth century, heat was widely—but not, in general, dogmatically—considered to be a material fluid, increasingly named "caloric." There were attacks on caloric in favor of "dynamical" theories of heat as "motion," but they seem to have had very limited impact (see below). Instead, the bulk of debate was concerned with *alternative* theories of caloric.

There were two main rival theories, one largely French in origin and support, the other British. Both saw heat as a material fluid, albeit extremely "subtle" (i.e., tenuous) and very probably "imponderable" (i.e., weightless), whose indestructibility guaranteed consistent calorimetric measurement. But there were important differences. The predominantly French theory of Lavoisier (and Laplace), on the one hand, saw "caloric" as a substance more or less on a par with the conventional chemical elements. According to Berthollet: "If one hesitates to regard this similarity between the properties of caloric and those of a substance entering into chemical combination as a conclusive proof of its [caloric's] materiality, one cannot but agree that the hypothesis that it exists presents no difficulties and has the advantage of involving only general and consistent principles in the explanation of phenomena" (Berthollet, 1803, i, p. 180; Fox, 1971, p. 124). Consequently, the Lavoisier "chemical" theory presumed that caloric could *either* be "free" and thus "sensible" (i.e., detectable as heat or change in temperature) *or* it could be "combined" with "ponderable" (i.e., having weight) matter and become "latent" (i.e., *not* detectable). So the latent heat of fusion of ice, say, involved a quantity of caloric combining with particles of ice in a chemical manner and thereby becoming "latent" and undetectable, just as oxygen might become undetectable in an oxide. On the other hand, the mainly British theory due to Irvine and Crawford rejected some of the implications of this "chemical" approach. Irvine proposed that there was no such thing as hidden, "latent" heat; any apparent appearance or disappearance of heat without a corresponding "sensible" change in temperature was explained as the result of changes or differences in heat capacities. Thus the latent heat of fusion of ice was to be explained by an alleged increase in the heat capacity of water as compared to ice. For Lavoisier and Laplace, addition of caloric could *cause* changes of state, especially from liquid to gas, just as combination with oxygen caused carbon to become gaseous carbon dioxide. For Irvine and Crawford an absorption or emission of heat simply *accompanied* the change of state. The same general mode of explanation was applied to the heat emitted in chemical reactions.

It was this conflict that provided the background for much of the early nineteenth-century research and debate on heat. The caloric theory of *gases*, in particular, based on the assumption of repulsive forces between gas molecules, lent itself very well to development within the Laplacian framework, and became one of the major successes of the program. As already recorded at the start of this section, in 1823 Laplace could claim that "...the phenomena of expansion, heat and vibrational motion in gases are explained in terms of attractive and repulsive forces which act only over insensible [very short] distances." But important developments also took place in areas relatively unconnected with the caloric theory, especially in relation to heat *transfer*. Overall, developments took place both in theoretical terms (e.g., in calculation of the speed of sound, or in the analytical theory of conduction) and experimentally (e.g., measurement of the specific heats of gases, or detection of radiant heat).

From 1815 onwards the caloric theory in any form became increasingly irrelevant, at least to many of the elite French scientists. But the caloric theory was not replaced by any alternative, "dynamical" theory; rather, there was a shift to a more agnostic or positivist view of the value of speculation about the fundamental nature of heat.

THE THERMAL PROPERTIES OF GASES

It seems therefore that general laws respecting the absolute quantity and nature of heat, are more likely to be derived from elastic fluids [gases] than from other substances.... (Dalton, 1802; Cardwell, 1971, p. 128)

In 1800 there was little reason to suppose that the thermal properties of gases (as distinct from solids or liquids) would be especially simple or revealing. True, Boyle's law was simple and apparently universal, but no uniformity had been discovered in the coefficients of expansion of different gases or in their specific heats. This situation changed very quickly in the first decade of the new century. By 1802 it was established that most gases expand more-or-less equally as their temperature increases, and by 1816 at the latest it was generally accepted that this expansion was linear. At the same time, important if not entirely conclusive steps were taken towards the accurate measurement of the specific heats of different gases. In the early 1820s, Laplace and Poisson were able to present comprehensive accounts of the behavior of gases, still largely based upon the caloric theory. Employing the newly available experimental data, the accuracy of their theory was confirmed by its precise prediction of the speed of sound.

The Thermal Expansion of Gases

At the end of the eighteenth century, although there had been little substantial development since the work of Amontons some 100 years earlier, interest in the thermal expansion of gases was stimulated by two distinct sets of problems. On the one hand, fashionable enthusiasm for mountaineering and ballooning directed attention towards the accurate measurement of altitude. This was commonly done with a barometer—the column of mercury falls fairly constantly by about 1 inch per 1,000 ft. (1 cm per 120 m) of ascent—but it was realized that results could be distorted by changes in temperature. Understanding exactly how changes in temperature affected air pressure would allow suitable corrections to be made. Better understanding of the thermal expansion of the air would also help astronomers to correct observational errors caused by the refraction of starlight entering the atmosphere. On the other hand, there were exciting developments in "pneumatic" chemistry, especially the rapid discovery of numerous new chemically distinct gases; this prompted study of their response to changes in temperature in order to standardize comparisons between them.

Even so, by 1800 few reliable results had been obtained. The most authoritative work, by the French chemist Guyton de Morveau (1737–1816) in 1789, suggested that different gases expanded to very different extents, and more rapidly at higher temperatures. These studies encouraged two other scientists to take up the problem in the first couple of years of the new century. One was English, the other French, and they worked independently. Although their work well illustrates the different national styles, they both produced similar results—although they did not necessarily draw similar conclusions from them.

John Dalton (1766–1844) is best known for his 1808 *New System of Chemical Philosophy* in which he developed his theory of chemical atomism, the idea that each chemical element is composed of identical atoms distinct from the atoms of every other element. Dalton's theory grew out of his work on the atmosphere and the diffusion of gases. It was in this context in 1801 that Dalton presented a paper to the Manchester Literary and Philosophical Society, "On the Expansion of Elastic Fluids by Heat," in which he described his studies of a range of gases, such as air, oxygen, hydrogen, nitric oxide, and carbon dioxide. He suspected, very probably correctly, that the presence of moisture had distorted the results of earlier experimenters, so he took particular care to dry his gases. Although he admitted that his results were less than perfect, he concluded that between room temperature and the boiling point of water all the gases tested expanded in the same proportion. His results suggested a coefficient of expansion equal to 1/256 per °C.

At the same time, but quite independently, the topic was being investigated by Gay-Lussac. Apart from the stimuli mentioned above, he was probably encouraged to undertake this research by Laplace, who particularly wanted to know how temperature variation might affect atmospheric refraction and distort astronomical observations. As might have been expected from one of France's new professional scientists, Gay-Lussac's "Researches on the expansion of gases and vapours" were much more extensive and systematic than Dalton's, and his report, running to 20 pages compared to Dalton's mere 4, gave a detailed description of his apparatus and methods. Among other things, he divided the gases studied into two groups, water soluble and insoluble, and performed his experiments on the latter over mercury. Nevertheless, Gay-Lussac obtained much the same results as Dalton, finding that all the gases tested expanded equally over the range 0 to 100°C, with equal coefficients of expansion—which he measured as 1/267 per °C.

Thus Dalton and Gay-Lussac agreed that all gases expanded equally for the rises in temperature that they had studied. At this point neither of them had attempted to measure the expansion of gases at intermediate temperatures to determine whether the expansion was uniform or *linear*. For various complex reasons connected to his commitment to the Irvine version of the caloric theory, Dalton rejected the reliability of the mercury thermometer and denied that the expansion of gases was truly linear. Gay-Lussac, on the other hand, suspected that the expansion probably *was* linear, so that the volume (V) at temperature (θ) could be expressed by the formula:

$$V = V_0(1 + \alpha\theta)$$

where V_0 was the volume a 0°C, and the coefficient of expansion, α, was 1/267. (This relationship is sometimes known, rather unfairly, as "Charles' law"; in his paper Gay-Lussac mentioned earlier unpublished experiments by "Citoyen Charles," who was none other than the pioneer balloonist J.A.C. Charles encountered above; Charles had concluded, in fact, that the law of equal expansion was *only* true for insoluble gases and *not* for soluble ones.)

Later unpublished studies conducted by Gay-Lussac tended to confirm his suspicions. It was only in 1817, however, that the problem was cleared up to most people's satisfaction by another two French scientists working together, Pierre Louis Dulong (1785–1838) and Alexis Thérèse Petit (1791–1820). They conducted further experiments on the expansion of various gases, along with a painstaking comparison of the relative expansions of the most widely used thermometric substances, namely, mercury and air. They concluded that the air thermometer provided the best available candidate for an absolute temperature scale and, somewhat over-enthusiastically, that "this confirms that Mariotte's [Boyle's] law is exactly true, whatever the temperature" (Dulong and Petit, 1818; Cardwell, 1971, p. 139). Thus Boyle's law and the laws of temperature variation were combined, almost by definition, and it was possible for Siméon-Denis Poisson (1781–1840) to state explicitly in 1823 that the pressure (P), density (ρ) and temperature were connected by the equation

$$P = a\,\rho\,(1 + \alpha\theta)$$

where "a" is a constant. Allowing for the absence of the absolute temperature scale, this equation is equivalent to the modern equation of state for an ideal gas, $PV = nRT$.

In retrospect, Gay-Lussac's law immediately suggests the existence of an absolute zero of temperature, at which the volume of a gas would shrink to nothing. Using Gay-Lussac's widely accepted value for α, 1/267 per °C, absolute zero would then be at –267°C. There had indeed been earlier speculation about the existence and value of an absolute zero of temperature, especially within the Irvine school. Dalton, for example, had calculated various figures in the region of –6,150°F (–3,430°C). More conventional calorists, however, were very cautious about such deductions because for them, having rejected the Irvine theory, the total absence of "sensible" caloric would still allow the presence of "latent" caloric, so there would be no absolute zero of *heat*. Gay-Lussac declared that "the determination of the absolute zero of heat must appear an utterly fanciful question" (Fox, 1971, p. 151).

Adiabatic Phenomena and the Speed of Sound

At the very start of the nineteenth century the importance of gases was further enhanced by investigation of "adiabatic" phenomena. A body of gas may get hotter or colder, without heat entering or leaving it, as a result of compression or expansion. (When the air in a bicycle pump is compressed it gets hotter, even though no heat has been transferred to it. If, when the tire has returned to room temperature, the [compressed] air inside is allowed to escape, it will emerge colder than room temperature, even though no heat has been removed from it. Such situations, in which no significant heat flow [in or out] takes place, either because of good insulation or because of the speed of the process, only came to be labeled "adiabatic" [from the Greek for "impassable"] in 1858.) References to adiabatic phenomena before the late eighteenth century are rare. The first person to notice such an effect seems to have been Robert Boyle in 1665. He noted that a thermometer in the evacuated "receiver" of a vacuum pump rose slightly when air was allowed to re-enter, and conversely that the temperature fell slightly as the air was pumped out. The effect was small and apparently trivial: It was attributed by Boyle to mechanical distortion of the thermometer bulb, and this, or possibly evaporation of residual moisture on the bulb, were the common explanations in subsequent years. Only in the last couple of decades of the eighteenth century did a more systematic scientific interest develop.

Once the reality of the effect was acknowledged, a variety of explanations was proposed. Adiabatic effects are largely due to the interconversion of external work and internal heat as a gas is compressed or expands; when this was recognized in the 1850s, adiabatic effects came to be seen in retrospect as powerful, even fatal, arguments against the caloric theory. In the early nineteenth century, however, this was simply not the case; a couple of quite plausible caloric-based explanations were available. For example, the early prominence of the vacuum pump in this context focused attention on the role of the vacuum itself; it was postulated that "empty" space would have a heat capacity and content in its own right—hence, when air re-entered the vacuum chamber, bringing its own heat with it, there would be an overall increase in the amount and density of caloric, and thus an increase in temperature. More commonly, the heating (or cooling) of a body of gas when compressed (or expanded) could be interpreted in terms of the heat fluid being squeezed out of (or sucked up, sponge-like) from the caloric atmosphere supposed to surround each atom of the gas. Given that by its nature heat was "expansive," it was simply presumed that gases needed a certain amount of heat, regarded as a "latent heat of expansion," to maintain their temperature as they expanded. The view of mineralogist René Just Haüy (1743–1822) was typical: "Whenever we define specific caloric—the caloric which is needed to raise the temperature of a substance by a given number of degrees—there is included, by our definition, that part of the whole whose sole function is to bring about expansion, this latter effect being a necessary accompaniment of a rise in temperature" (Haüy, 1806, vol. 1, p. 118; Fox, 1971, p. 160).

From 1800 onwards, interest in adiabatic phenomena quickened noticeably, especially within the French scientific community. This was prompted by two important developments, the first of which was the invention of the "fire-piston." The fire-piston was the result of the accidental discovery in 1802 that the very rapid compression of air in a tube by a tightly fitting piston could cause heating so intense that a piece of tinder in the bottom of the tube would catch fire. This promptly resulted in a commercial product that, for a couple of decades, until the introduction of modern phosphorus matches, competed with the traditional tinder-box. More profoundly, however, it demonstrated vividly that the heating of a gas by compression was not just some slight and possibly spurious effect, but could be very substantial.

The second important development was the realization that taking account of adiabatic heating and cooling might allow an improved estimate of the speed of sound. In 1800 it had long been known that the speed of sound as calculated according to reliable Newtonian principles was significantly low in comparison with experiment. Newton himself had shown in the *Principia* that the speed of waves in any medium depends upon the elasticity and (inversely) upon the density of the medium. For air, however, this resulted in a theoretical estimate of the speed of sound as about 970 feet per second (300 m/s), whereas the best available experimental measurements were around 1,100 feet per second (330 m/s).

There was promising, but partial, early success in resolving this difference by Biot in 1802. According to Biot, Laplace had asked him to investigate the problem, with a view to taking into account the possible adiabatic heating and cooling of the air, on the assumption that the rapid compression and expansion of the air in a sound wave would not allow time for the diffusion of signifi-cant amounts of heat. Exploiting Gay-Lussac's recent work on the expansion of gases with increase in temperature, Biot made plausible (albeit incorrect) assumptions about the increase of temperature, and hence pressure, that might result from a decrease in volume under adiabatic conditions. Reworking the Newtonian calculations on this basis resulted in an estimated speed of sound of 1,362 ft./sec. This was clearly much too high, but was taken to confirm at least that some temperature changes did occur in sound waves. The eventual triumphantly systematic and accurate—and caloric-based —treatment of the topic presented in the early 1820s (see below) depended upon the determined investigation of the specific heats of gases in the intervening years.

The 1812 Competition and the Specific Heats of Gases

In the two decades between 1805 and 1825 there were substantial develop-ments in the understanding of the specific heats of gases. These developments, which were both experimental and theoretical, were given focus and incentive by a series of prize competitions set by the First Class of the Institute, starting in 1812. As a result, from 1813 onwards accurate measurements of the specific heats at constant pressure of various common gases became available, based on increasingly sophisticated experimental apparatus and procedures. From 1816 there emerged an explicit recognition of the importance of the distinction between specific heats measured at constant pressure and at constant volume; in the early 1820s good experimental measurements of the ratio between the two became available, which enabled the accurate derivation of the speed of sound. Initially, at least, these developments served to further consolidate the validity and vitality of the caloric theory in the form proposed by Lavoisier and Laplace.

Early stimuli to the program of (predominantly French) investigations were various. The basic challenge, as specified in the terms of the Institute com-petition, was to "determine the specific heats of gases, in particular those of oxygen, hydrogen, azote [nitrogen], and some compound gases, comparing them with the specific heat of water" (Fox, 1971, p. 134). An explicit expectation was "that the determination of the specific heats of gases will lead to the solu-tion of the outstanding problem of whether some caloric exists in substances in a combined state or whether all the heat given out in a reaction is due to a change in the specific heat of the reacting substance," to a resolution, in other words, of the Lavoisier versus Irvine controversy—hopefully in favor of the French. But the recently highlighted adiabatic phenomena were also an issue.

The intuitively satisfying "sponge" model of adiabatic heating and cooling remained relatively vague, but seemed to imply that the specific heat capacity (by weight) of a fixed body of gas should *decrease* as its density was increased by compression (although in fact it remains more-or-less constant), thereby accounting for the liberation of excess heat upon compression. Hence a second requirement of the Institute prize was to "determine, at least approximately, the change in specific heat that is produced when the gases expand..." (ibid.).

Work on these problems predated the Institute competition. Between 1807 and 1812 Gay-Lussac conducted a variety of often ingenious but sometimes rather crude experiments. For example, he investigated the alleged heat capacity of a vacuum by enclosing a sensitive thermometer in the Torricellian vacuum at the top of a mercury barometer: No change in temperature was observed when the vacuum was "compressed" or "expanded" by tipping the tube. Another series of experiments involved a pair of large glass flasks connected by a tube; one of the flasks having been evacuated, the various gases in the other were allowed to enter and the changes in temperature noted, from which relative specific heats were calculated. In 1812 he reported upon a new series of experiments involving a rather crude method of mixtures, but the overall results were confused and contradictory. The Institute competition was intended to produce definitive experimental answers to the questions that Gay-Lussac had tackled rather inconclusively. It was very successful in this, although there were only two entries, each from a pair of researchers working together. Both the entries were very competent experimentally, producing similar and accurate measures of the specific heats of a range of gases. The work of Nicolas Clément (1778 or 79–1841) and Charles Bernard Desormes (1777–1820), which lost partly because of its unorthodox theoretical framework, will be

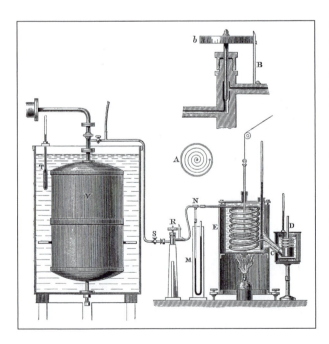

Figure 2.2: The apparatus with which Regnault conducted a definitive series of measurements of the specific heats of gases in the 1840s, a refined version of the apparatus used by Delaroche and Bérard in 1813. The gases under examination were heated in the spiral tube (between N and C) and then allowed to flow through the calorimeter at D. Preston, T., *Theory of Heat* (London, 1904), fig. 68.

discussed briefly below. It was the work of François Delaroche (c.1780?–1813) and Jacques Etienne Bérard (1789–1869) that won the prize, and this will now be discussed in more detail.

Delaroche and Bérard ran two main series of experiments. In the first series, to measure the specific heats of a range of gases, they chose to measure specific heat at constant pressure, largely because it was easier in practice. They adopted a theoretically unremarkable constant flow technique. Careful consideration was devoted to the elimination of heat losses by radiation, conduction, and convection. Two variations of this basic method produced reassuringly similar sets of results for air, oxygen, hydrogen, carbon monoxide, nitrous oxide, carbon dioxide, and ethylene. Several conclusions were evident. For one thing, it seemed clear that the specific heats by volume of all gases were *not* all exactly the same, as had been suggested. More importantly for Delaroche and Bérard, and for the Lavoisier and Laplace theory of combined caloric, the specific heats of some of the gases appeared to be quite incompatible with the Irvine theory: "[W]e must reject once and for all the hypothesis whereby the release of heat which occurs when substances combine is attributed solely to a change in the capacity of these substances for caloric, and consequently we must accept, with Blake [*sic*, i.e., Black], MM. Lavoisier and De Laplace and a large number of physicists, that caloric exists in a combined state in bodies" (Delaroche and Bérard, 1813, pp. 174–5; Fox, 1971, p. 142). The Irvine theory was never again a serious contender.

The second part of the competition, which Delaroche and Bérard tackled in an additional experiment, was aimed at measuring the specific heat of air at greater than atmospheric pressure and density. Their admittedly rough results indicated—as was widely expected at the time—that the specific heat of air did indeed *decrease* as the density increased.

> The specific heat of atmospheric air…considered in the ratio of its mass… diminishes as the density increases.…Everyone knows that when air is compressed heat is disengaged. This phenomenon has long been explained by the change supposed to take place in its specific heat; but the explanation was founded upon mere supposition, without any direct proof. The experiments which we have carried out seem to us sufficient to remove all doubts upon the subject. (Delaroche and Bérard, 1813; Cardwell, 1971, p. 136–7)

Expectation, combined with the authority of their victory in the Institute competition, thus consolidated the caloric understanding of adiabatic effects. This mistake has been described as "one of the most significant ones ever made in the history of science" (Cardwell, 1971, p. 137).

Specific Heats and the Speed of Sound

Delaroche and Bérard had measured specific heats at constant pressure (c_p) rather than at constant volume (c_v), largely as a matter of practical convenience. A clear assertion of the distinctive importance of *both* specific heats was first made by Laplace in 1816. His discussion of specific heats occurred in a paper on "The speed of sound in air and water," which claimed that the result of taking adiabatic effects into account was to introduce a correction factor (c_p/c_v)—now usually labeled γ (gamma)—into the Newtonian equation. How Laplace arrived at this adjustment was only explained in a later series of papers published in the early 1820s. Here he developed his results within a complex elaboration of the traditional caloric model of gases: The repulsive force between any two particles of gas was assumed to depend upon the quantity of caloric contained in each particle. As one would expect from Laplace, the model was given a rigorous mathematical development; with the addition of a further precise but somewhat arbitrary assumption that some of the heat "expelled" by compression remained latent, the required equations were duly generated, as was, in combination with the experimental determinations of γ, a very satisfactory estimate of the speed of sound.

In 1823 Poisson reworked his master's material, reaching similar results but by a more straightforwardly descriptive approach that avoided Laplace's complex hypothetical mechanisms of caloric distribution. Poisson based his analysis on a closed three-stage cycle of changes to a fixed body of gas: In the first stage the gas was supposed to be expanded by heating at constant pressure; in the second stage adiabatic compression (and consequent further heating) returned the gas to its original volume; finally, the gas was allowed to cool to its original temperature and starting point, thereby completing a closed cycle of changes. Poisson could thereby link c_p involved in the first stage with adiabatic expansion in the second stage and c_v involved in the third. Combined with the laws of Boyle and Gay-Lussac, this allowed him to derive the laws governing the behavior of gases undergoing adiabatic changes, especially that pV^γ is constant. According to Fox (1971, p. 177), "Poisson's greatest contribution, then, was rather to free Laplace's work of its more suspect elements…showing in a most effective manner just how irrelevant much of Laplace's theory was, so that even to a reader convinced of the physical reality of caloric Poisson's must have seemed undeniably the more fruitful approach."

These theoretical developments elicited an experimental response, both in measurement of the speed of sound and in determination of c_v or γ. At Laplace's instigation, new measurements of the speed of sound were made, with little resultant adjustment to the accepted value. Direct determination of c_v was difficult for practical reasons, however. It was at this point that the unsuccessful entry for the 1812 competition by Clément and Desormes finally came into its own, despite its unconventional, indeed old-fashioned, theoretical framework. Because a thermometer suspended in a vacuum still registers a temperature, they had maintained that the vacuum had a definite temperature and heat capacity of its own, a theory that had been quite popular in the eighteenth century. Their original experiment, therefore, was essentially the method of mixtures applied to vacuum "mixed" with air or other gases. Laplace argued that their experiments could be reinterpreted as a cycle of isothermal, adiabatic, and iso-volumetric changes (similar to that constructed by Poisson), from the results of which it was possible to derive a value for γ, and it is in this role that their apparatus is remembered today. From their results Laplace derived a value of 1.35 for air; Gay-Lussac repeated the crucial experiments and obtained values around 1.373, which was close to the modern value of 1.40 and implied a speed of sound of 336 m/s.

In the later 1820s Dulong went on to suggest on experimental grounds that, other things being equal, a given decrease in the volume of any gas resulted in the release of the same amount of heat, which is more or less equivalent to the modern assertion that $(c_p–c_v)$ is constant. A generation later this would be seen as powerful evidence for the mechanical origin of adiabatic heating effects. At the time this was not at all apparent, however. Even so, and despite its achievements both theoretical and experimental, the caloric theory was by now in terminal decline, as will be discussed in more detail below.

THE TRANSFER OF HEAT

... a very extensive class of phenomena exists, not produced by mechanical forces, but resulting simply from the presence and accumulation of heat. This part of natural philosophy cannot be connected with dynamical theories, it has principles peculiar to itself, and is found on a method similar to that of the other exact sciences.... (Fourier, 1955 [1822], p. 23; Cardwell, 1971, p. 118)

During the eighteenth century little further attention had been paid to processes of heat transfer since Newton's discussion of cooling. The chemical context of the caloric theory tended to focus attention on the static distribution of the material fluid rather than its "flow." This emphasis changed after 1800. In 1804, John Leslie (1766–1832), professor of mathematics at Edinburgh, clearly distinguished the various processes involved in heat transfer, namely, radiation, conduction, and convection—albeit under the names "pulsation," "abduction," and "recession." In the following quarter-century the precise nature and properties of these different mechanisms were gradually analyzed, especially radiation and conduction.

Calorific Radiation or Radiant Heat

The distinctive nature of "radiant" heat was recognized in the late eighteenth century, and its similarity to visible light was gradually confirmed. The marked difference between radiant heat and the heat rising above a stove was clearly argued by the Swedish apothecary Carl Wilhelm Scheele (1742–86) in his *On air and fire* in 1777. Scheele pointed out that radiated heat did not necessarily warm the intervening air, nor was it deflected by cold draughts. Like light, it traveled in straight lines and could be reflected, although, unlike light, radiant heat was cut off by glass, a fact that was to cause considerable initial confusion.

In 1800 the astronomer Frederick William Herschel (1738–1822), who in 1781 had discovered the new planet Uranus, identified what came to be known as infrared radiation. Investigating the heating effect of different parts of the visible spectrum from the sun, Herschel was surprised to find that a heating effect was detectable, indeed strongest, *beyond* the red end of the spectrum. In the following early decades of the nineteenth century it was gradually established that this invisible "calorific radiation" shared the properties of reflection and refraction (and, eventually, polarization) with ordinary visible light. Nonetheless there were obstacles to regarding radiant heat and light as essentially identical; as noted by Scheele, radiant heat was cut off by glass even though visible light was transmitted. Increasingly refined researches by Delaroche in 1811, by Bérard in 1813, by the German Thomas Johann Seebeck (1770-1831) in 1818–19, and most convincingly by Macedonio Melloni (1798–1854) in the 1830s demonstrated that different materials would absorb different parts of the thermal spectrum to differing extents, just as colored glass absorbed different parts of the visible spectrum. Melloni had the advantage of using a "thermopile," based on the thermoelectric effect discovered by Seebeck in 1826: A dozen or more paired metal strips or "thermocouples," usually of the metals antimony and bismuth, arranged in series provided a highly sensitive heat detector. With this device Melloni established that rock salt was "diathermanous," that is, more or less transparent to heat radiated from bodies at all temperatures.

Thus it came to be accepted that calorific rays were of the same nature as light, and that radiant heat and visible light formed part of a continuous spectrum of radiation. "In place of this complication of ideas," said Biot (1816, iv, p. 651; Cardwell, 1971, p. 112), "let us conceive simply and conformably to the phenomena that solar light be composed of an ensemble of rays, unequally refrangible [refractable]...which supposes original differences in their (molecular) masses, speeds and affinities." This only partly visible spectrum of radiation was further extended in the same period with the discovery of chemical effects beyond the violet end of the spectrum. The subsequent history of radiant heat is charted in chapter 7.

In the early nineteenth century, following Newton, light was very widely regarded as a material, corpuscular phenomenon. Hence Biot's reference to the "molecular masses" of rays of light. The identity of radiant heat and light was thus readily accommodated to a material, caloric theory of heat; calorific rays were simply the emanation of very fine particles of caloric. From about 1815 onwards, however, the alternative wave theory of light became increasingly popular, thanks to the efforts of Thomas Young (1773–1829), Fresnel, and

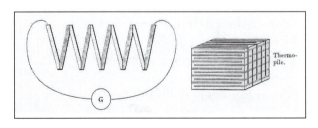

Figure 2.3: The thermopile, invented by Macedonio Melloni in the 1830s to detect and measure radiant heat, consisted of an array of sensitive thermocouples. Preston, T., *Theory of Heat* (London, 1904), fig. 165 and 164.

others; within this framework radiant heat conceived as a wave or vibration helped to undermine the caloric theory.

Cooling, Conduction, and Cosmology

It had always been evident that different substances transmitted heat more or less readily—hence the wooden handle of an iron frying pan. The concept of thermal conductivity gained extra credibility from electricity, where the distinction between conductors and insulators was more conspicuous, although initial studies tended to confuse the roles of specific heat capacities and conductivities. In the late 1780s, for example, the Dutch botanist Jan Ingen-Housz (1730–99) dipped identical rods of different materials in wax; when the wax had set, the ends of the rods were simultaneously immersed in boiling water, and the different speeds at which the boundary of melting wax moved were observed.

In 1804 Biot conducted experimental and mathematical studies of the distribution of temperature along an iron bar heated at one end. This helped to stimulate the definitive treatment of heat transfer in general and conduction in particular by Fourier; initially presented in the form of the winning entry to the 1811 Institute physics competition, this treatment was belatedly published in 1822 as *The analytical theory of heat*. The delay in the wider publication of Fourier's ideas probably owed something to the opposition of the French scientific establishment. Having narrowly escaped the guillotine in 1794, Fourier accompanied Napoleon on his military campaign to Egypt in 1798–9; on his return to France, however, he was consigned to provincial administration, only returning to Paris after Napoleon's downfall in 1815. Fourier's basic approach to science was radically at odds with the Laplacian program. Fourier had little time for speculative hypotheses involving unmeasurable forces between invisible atoms, or for subtle fluids. He aimed rather to construct a quantitative, but purely descriptive, analysis of thermal phenomena, an axiomatic, mathematical theory of heat comparable to Newtonian rational mechanics.

According to Fourier, the thermal behavior of bodies was determined by three properties: firstly, their specific heats; secondly, their "exterior conductivities";

Figure 2.4: The apparatus used in 1822 by Belgian physicist César Despretz (1798–1863) to measure the conduction of heat along an iron bar. Preston, T., *Theory of Heat* (London, 1904), fig. 193.

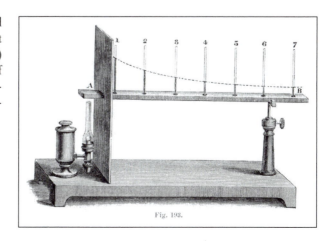

Fig. 193.

and lastly, their "interior conductivities." Specific heat was already well understood. Exterior conductivity, the rate at which a body lost heat from its exterior surfaces, involving the combined effects of radiation, conduction, and convection, remained complex and largely intractable, although Fourier provided a pragmatic, experimental definition. With the third property, interior conductivity (K), Fourier was more enduringly successful; for any given substance it was defined in terms of a thin parallel-sided sample and the quantity of heat that "flows during one minute across a surface of one square metre taken on a section parallel to the extreme planes supposing that these planes are maintained, one at the temperature of boiling water, the other at the temperature of melting ice and that all the intermediate planes have acquired and retain a permanent temperature" (Fourier, 1955 [1822], p. 52; Cardwell, 1971, p. 117). Such a definition had two virtues. On the one hand, it suggested precisely the kind of accurate experimental measurement at which the professional French scientists increasingly excelled. On the other hand, if K was found or assumed to be more or less constant, then the definition could be translated into a neat differential equation asserting that the rate of flow of heat (dQ/dt in calculus terms) was directly proportional to the conductivity, the area (A) across which it flowed, and the temperature gradient (dθ/dx), namely:

$$dQ/dt = -KA.d\theta/dx$$

Combined with definitions of the other variables, Fourier could on this basis construct and solve sets of differential equations for the distribution of heat in space and time throughout bodies of different shapes (e.g., cylinders, spheres) and for groups of bodies. In particular, he analyzed the distribution of heat within the solar system, between the hot sun, the Earth, and cold space. Studies of this kind continued to play a role in cosmological-theological debate throughout the rest of the nineteenth century.

The difficulty of analyzing *exterior* conductivity was soon demonstrated experimentally. In 1817, stimulated by Fourier's theorizing and by long-standing doubts about the validity of Newton's law of cooling, the Institute announced a competition to study empirically the actual laws of cooling, both *in vacuo* and in a range of gases. The challenge was taken up (and the prize won) by Dulong and Petit, fresh from their work on gases and temperature; the laws of cooling that they found were very complex, even in the case of cooling *in vacuo*, which largely depended simply on radiation. Nevertheless, in 1819 they were able to use their measurements as a means of measuring the specific heat of 19 chemical elements. Quite by accident, apparently, they noticed that in many cases the specific heats (per unit mass) were inversely proportional to the atomic weights of the elements; this suggested that there was a more-or-less fixed specific heat *per atom* or, in modern terminology, that molar specific heats were constant for a wide range of elements—the Dulong and Petit law of atomic heats. This discovery provided an independent method for resolving the vexed problem of estimating atomic weights and at the same time lent support to Dalton's still highly contentious atomic theory. Indirectly, this also served to undermine the caloric theory.

THE NATURE OF HEAT

> [I]t appears to me to be extremely difficult, if not quite impossible, to form any distinct idea of any thing, capable of being excited, and communicated, in the manner the heat was excited and communicated in these experiments, except it be MOTION. (Thomson, B., 1798, p. 99)

In 1783 Lavoisier and Laplace had side-stepped the competing claims of the material "caloric" and the dynamical "heat is motion" theories of heat. Nevertheless, during the Napoleonic era, the caloric theory was very widely accepted; although in general this acceptance was neither uncritical nor dogmatic, outright rejection of caloric was very rare. The most conspicuous critics of caloric were a set of natural philosophers active in London around the turn of the century, although their attacks had very limited impact. Even so, after 1815 enthusiasm for caloric began to wane, even in France, as the result of a combination of scientific and institutional factors.

"A Shapeless Hypothesis": The Fortunes of the Dynamical Theory

The buccaneering career of Benjamin Thomson, Count Rumford (1753–1814), (partly funded by marrying rich widows, including lastly, unhappily, and rather ironically, Mme Lavoisier) took him from revolutionary America to England, to Bavaria (in southern Germany), and to France. Towards the very end of the eighteenth century he was instrumental in founding the Royal Institution in London, but from 1784 to 1798 he served as minister for war in Bavaria; in this role he not surprisingly became involved with munitions and published papers on projectiles and gunpowder. Rumford's modern reputation rests largely upon his paper, "An Inquiry concerning the Source of the Heat which is Excited by Friction," presented to the Royal Society in 1798.

He recorded that "being engaged, lately, in superintending the boring of cannon, in the workshops of the military arsenal at Munich, I was struck with the very considerable degree of heat which a brass gun acquires, in a short time, in being bored; and with the still more intense heat...of the metallic chips separated from it by the borer" (ibid., p. 81). By immersing one end of the cannon barrel in a tank of water and using a deliberately blunted borer, he showed that in less than two-and-a-half hours the water could be brought to the boil merely by frictional heat. Rumford attempted to undermine any caloric explanations that might be offered; his main, and very powerful, argument was quite simply that the supply of heat seemed to be limitless: "In reasoning on this subject, we must not forget to consider that most remarkable circumstance, that the source of the heat generated by friction, in these experiments, appeared to be *inex-*

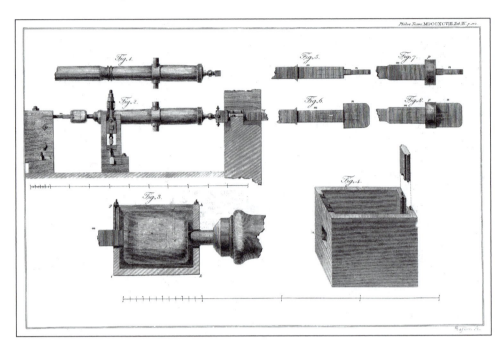

Figure 2.5: The cannon-boring machinery used by Benjamin Thomson, Count Rumford, in his 1798 "Inquiry concerning the Source of the Heat which is Excited by Friction." The rough cast brass cannon as it came from the foundry (fig. 1) was fixed in machinery (fig. 2) that rotated it against a blunt iron borer (m); in fig. 3 the stub end of the cast was contained in a wooden box (g, h, i, k) filled with water, which was heated by the friction. *Philosophical Transactions of the Royal Society*, vol. 88; by permission of the Syndics of Cambridge University Library.

haustible. It is hardly necessary to add, that any thing which any *insulated* body, or system of bodies, can continue to furnish *without limitation,* cannot possibly be a *material substance....*" (Thomson, B., 1798, p. 99). Instead, therefore, heat must be a form of motion. A few years later Rumford's argument was supported by an experiment described by the young chemist (eventually, Sir) Humphrey Davy (1778–1829), in which two blocks of ice allegedly melted when rubbed together *in vacuo* by a clockwork mechanism.

It was easy to criticize Rumford's and Davy's experiments and their arguments. It was well known, for example, that an apparently endless supply of electric charge could be generated by the rubbing of amber or glass with suitable cloths, yet no one doubted that electricity was some kind of fluid. Nevertheless, it was widely conceded that the generation of heat by friction was a weak point in material, caloric theories of heat. For most of Rumford's contemporaries, however, to abandon the caloric theory would have been very unrewarding. His assertion that heat was motion remained a fairly vague, qualitative analogy, unable to offer very satisfactory, let alone quantitative, explanations for latent heat, heats of chemical reaction, or (before 1815) radiant heat. Later, in 1812, Davy proposed a more complex (but altogether hypothetical) theory attributing heat capacity and temperature to the vibration and rotation of the particles of ordinary matter. According to Leslie, however, the dynamical theory was "...a shapeless hypothesis...[that] explains nothing; it throws out a delusive gleam and then leaves us in tenfold darkness" (Leslie, 1804, p. 140; Cardwell, 1971, p. 107).

But most important was the widespread perception that acceptance of the dynamical theory, and the notion that somehow heat could be "generated," would completely undermine the assumption of the *conservation* of heat. Despite their professed agnosticism concerning the true nature of heat, Lavoisier and Laplace assumed that *both* theories entailed conservation, for this was fundamental to any calorimetric procedures. Deny this and the painstaking experimental foundations of much of the heat studies of the day would be annihilated. It was precisely this objection that initially caused William Thomson to resist James Joule's findings.

The Decline of the Caloric Theory

The caloric theory, dominant, even triumphant, in 1815, fell rapidly from favor in the following decade. This decline owed relatively little to specific experimental developments and, as we have just seen, cannot be attributed to the appearance of a more convincing alternative theory of the nature of heat. The eclipse of caloric can only really be understood within the context of the wider contemporary reaction against the Laplace-Berthollet paradigm, a reaction that was both scientific and institutional.

From the strictly scientific point of view this reaction involved the rejection of several important planks of the Laplace-Berthollet platform. From 1815 onwards the corpuscular theory of light was steadily supplanted by the wave theory, and Berthollet's chemical ideas were marginalized by increasing acceptance of the constant composition of chemical compounds and Daltonian atomic theory. These scientific developments were accompanied by institutional changes within the French scientific establishment. Laplace and Berthollet had dominated the scientific scene during the Napoleonic era partly through their control of institutions, journals, laboratory facilities, teaching appointments, and so on. The success of Fourier's conspicuously non-Laplacian treatment of heat conduction in the Institute's 1811 physics prize competition suggests an early crack in the façade, but otherwise the revolt against their authority began quite suddenly in 1815 with the debate on the nature of light that was precipitated by Fresnel's first paper on diffraction. The coincidence with the downfall of Napoleon—the battle of Waterloo and Napoleon's final defeat took place in June of that year—and the restoration of the Bourbon monarchy is not entirely accidental, for there was at the same time a widespread reaction against science in general and Laplace's person and authority in particular. One of Laplace's supporters later recalled that "[A]ll those who attacked the results contained in the *Mécanique Céleste* were hailed, and the entire liberal press was mobilized against the distinguished men of our past, who, it was said, were no more than idols that had to be destroyed" (Fox, 1971, p. 235). Consequently there was a gradual shift of control—the editorship of journals, membership of the Académie, and so on—into the hands of a new and predominantly younger group of scientists including Fourier, the physicist Dominique Arago (1786–1853), Fresnel, Dulong, and Petit, who were far less committed to the Laplace program.

Initially the nature of heat was not significantly at issue. The first direct attack on the caloric theory from within the Parisian scientific establishment came from Dulong and Petit in 1819 when announcing their law of atomic heats. This discovery provided powerful evidence in favor of Daltonian atomism and triggered a more general rethink on their part. Nor were they alone. Only four years later Laplace's definitive version of the caloric theory of gases in the *Mécanique Céleste*, vol. 5, was received in deafening silence. For the majority of elite scientists, it seems, the caloric theory had become more or less irrelevant; the complex, speculative, corpuscular systems devised by Laplace appeared sterile; far more fruitful were the purely descriptive but exact mathematical theories formulated by Fourier or Poisson, or the systematic accumulation of "positive" factual knowledge through the painstaking and accurate experimental measurements of Dulong and others.

There had been two specific developments that made this rejection of (or agnosticism about) caloric more sustainable. In the first place, developments in electrochemistry offered a powerful alternative to the caloric theory's explanation of heats of reaction—always one of its strong points. In the second place, a dynamic or vibrational view of heat would fit well with the new wave theory of light. If radiant heat were a form of light, and if it were now accepted that light in general was a wave phenomenon, then this suggested that heat as a source of such waves would also be some kind of vibration. Thus Cambridge professor Philip Kelland (1808–79): "If, as there is every reason to believe, light is the result to vibratory motion, we are naturally driven to enquire into the possibility of applying the same consideration to heat" (Kelland, 1837, p. 124; Cardwell, 1971, p. 114).

The rejection of caloric was most thoroughgoing among elite French scientists. It retained its hold longest among peripheral figures, such as Carnot (see chapter 3), and outside France. It also survived in textbooks well into the 1840s; even when an author was noncommittal about "the nature of heat," he might well continue to find the language of material fluids useful for teaching purposes. It is worth repeating that the caloric theory could not be rejected in favor of the dynamical or any other theory, which, as we have seen above, failed to provide adequate alternative explanations in many areas. This remained the case into the 1840s. The decline of the caloric theory parallels a conspicuous shift from chemistry to physics as the center of gravity of interest in heat; chemists became (if only temporarily—see chapter 7) uninterested in heats of reaction, while physical connections between heat and light and electricity became important. It was the neglected connections between heat and mechanics that turned out to be crucial in the following quarter-century.

THE MOTIVE POWER OF FIRE: SADI CARNOT AND THE ORIGINS OF THERMODYNAMICS

3

INTRODUCTION

The study of these [heat] engines is of the utmost interest. Their importance is immense, and their use increasing daily. They seem destined to bring about a great revolution in the civilized world.... If one day the heat engine is so perfected that it becomes cheap to erect and economical to run, it will combine all the qualities we could wish, and advance industrial technique to an extent that can scarcely be predicted. (Carnot, 1986 [1824], p. 61)

Introduction: Economy and Efficiency

In 1815, immediately after the end of the Napoleonic wars, English engineer-entrepreneur Humphrey Edwards introduced the latest steam engines into France. Only recently developed by his erstwhile business partner Arthur Woolf (1776–1837), the engines were of a radical new design: They had twin or "compound" cylinders and were driven by high-pressure steam. In England they had already recorded "duties" of over 50 million foot-pounds per bushel of coal, a considerable improvement in efficiency compared with the best Watt engines available, and far outstripping the old-fashioned Newcomen engines that were still widely used in France. For the preceding quarter-century, thanks to the Napoleonic wars, France had been largely isolated from Britain in commercial and technological terms, while Britain had been transformed by an industrial revolution increasingly based upon steam power. In the aftermath of eventual military defeat, the French were eager to catch up with the new technology. Edwards had installed 15 Woolf engines by the end of November 1817, and some 300 by 1824. In the interest of the national economy, however, the French professional scientific elite were equally keen to *understand* the reasons for the superior performance of the new engines. From the resulting debate there emerged an extraordinary—if initially quite neglected—contribution to the foundation of the modern science of thermodynamics.

In many ways the quarter-century from the early 1820s until the mid 1840s could be regarded as a quiet, even dull, period in the history of heat. The attention of the scientific community was largely focused on exciting developments in electromagnetism, chemistry, and light. Certainly, painstaking and ultimately invaluable experimental work on heat continued at the hands of Dulong and others. But it does seem that the protracted yet inconclusive debates in the first quarter of the nineteenth century about caloric and the nature of heat had undermined enthusiasm for further theoretical speculation.

However, there were conceptual developments almost inconspicuous at the time that would eventually be most significant. If it makes any sense to speak of the "birth" of thermodynamics in the 1850s (see chapter 5), then its "conception" would have to be located in the publication of *Reflexions on the motive power of fire and upon machines designed to develop that power* by a retiring and unregarded French engineer, Sadi Carnot (1796–1832), in 1824. Carnot's slim volume provided the first coherent articulation of many of the ideas, methods, and conclusions that were to be fundamental to the new science. Politely reviewed upon its first appearance, it was subsequently almost entirely ignored until after Carnot's death in 1832. The basic ideas were reworked and republished by another professional French engineer, Émile Clapeyron (1799–1864) in 1834, only to be similarly neglected for another decade.

In the mid-1840s the potential of the Carnot-Clapeyron approach was finally recognized by the brilliant young Scottish mathematical physicist William Thomson (later Lord Kelvin) (1824–1907); in 1848 he used it to construct the first *absolute* temperature scale, a scale that was independent of the particular properties of any specific material substances such as mercury or air. Thomson's career will be discussed at length below. When integrated with the work of James Joule and others on "the mechanical equivalent of heat" (see chapter 4), the new approach allowed the German physicist Rudolph Clausius (1822–1888), Thomson, and others to construct the new science of thermodynamics from 1850 onwards, as we shall see in chapter 5.

This later development took place very much within the realm of the new discipline of mathematical physics. Carnot, however, had been more of an engineer than an academic physicist, and his work was intended as a contribution to the growing literature on the theory and practice of heat engines. Thus his unique contribution is only comprehensible within the context of the practical and theoretical development of power technologies—both steam and water— in the late eighteenth and early nineteenth centuries.

Power Technology in the Late Eighteenth and Early Nineteenth Centuries

By 1780 the steam or "fire" engine was a well-established feature of the industrial landscape, especially in Great Britain. Power technologies—both steam and water—continued to develop rapidly after 1780 and provided a range of intriguing and important technical problems to be solved. At the same time they provided a growing body of increasingly skilled and educated engineers to tackle these problems, and thus there were articulated, through the late eighteenth and early nineteenth centuries, many of the concepts and much of the terminology and the empirical evidence involved in the construction of thermodynamics. After a quick look at some crucial technical developments, we shall look at the accompanying conceptual changes.

In 1775 Matthew Boulton (1728–1809), Watt's astute business partner, had contrived to have Watt's 1767 patent for the separate condenser extended for an extra 25 years to 1800. Although they were by no means the only manufacturers of steam engines, it is possible that Boulton and Watt's control of this key improvement inhibited further technical development in the last quarter of the eighteenth century. In particular, Watt's engines operated at only slightly above atmospheric pressure—i.e., at one or two pounds per square inch (psi) above the atmospheric pressure of about 15 psi. Watt opposed the introduction of high-pressure steam, mainly because of the very real danger of boiler explosion, but maybe also because a high-pressure engine could operate efficiently *without* his separate (and patented) condenser.

Be that as it may, high-pressure engines began to appear and spread rapidly after 1800. The crucial development was indeed a new type of boiler that could reliably withstand higher pressures. Another West Country engineering genius, Richard Trevithick (1771–1833), developed cast-iron boilers that he operated at up to 10 atmospheres, but more usually at 2 or 3 atmospheres. Similar developments took place in America at the hands of Oliver Evans (1755–1819). The high-pressure engine had several advantages. As Watt was aware, operation at high pressure meant that a steam engine could efficiently dispense with his separate condenser, simply expelling exhaust steam to the atmosphere. This in turn meant that such an engine could be smaller, simpler, and thus cheaper.

A high-pressure engine could also be operated "expansively." In Watt's beam engines the active "working" down-stroke was powered by low-pressure steam, injected above the piston and acting against the negligible resistance of the partial vacuum created below the piston by the condenser. Watt had realized that, if he cut off the steam supply above the piston early in the downstroke, the steam would still expand and continue to act upon the piston, albeit with decreasing pressure: At some loss of power he could make a substantial saving in fuel. For Watt's low-pressure engines this procedure turned out not to be effective in practice. In the new high-pressure engines, however, it proved to be very effective and became common practice. It was widely accepted that the new high-pressure engines were more efficient than low-pressure engines. This stimulated considerable discussion, both in England from the turn of the century and in France from 1815, when they were introduced there in the aftermath of the Napoleonic wars.

It should not be overlooked that the eighteenth and indeed the early nineteenth centuries also saw continued, steady development of *water* power. Even in 1830, when the use of steam power had expanded some 13 times since 1760, the use of water power had also more than doubled, and water still equaled steam in the provision of industrial power in Britain. Apart from being intrinsically rotational, water power could still be more powerful than steam. On the one hand there was a steady improvement in the efficiency of traditional water-wheels. This is particularly evident in English engineer John Smeaton's (1724–92) studies of the relative efficiency of over- and under-shot wheels, which are discussed below. On the other hand there was a lively exploration of alternative mechanisms for exploiting water power. Apart from a variety of turbines—which did not come into their own until the mid-nineteenth century—it is worth noting the "column-of-water" engine; this latter, as is explained in detail by Cardwell (1971, chapter 3), was explicitly modeled on the *steam* engine, but used a head of water to produce pressurized water, which then replaced steam as the driving force.

CONCEPTS OF WORK AND EFFICIENCY

I sell here, Sir, what all the world desires to have—Power. (Matthew Boulton, letter to James Boswell [1740–95], March 22, 1776. [Boswell, *Life of Johnson*, vol. 2, p. 459])

Introduction

Although the practical, technical details of steam engine design and operation were increasingly well documented—more effectively by French writers than by British, in fact—theoretical understanding remained undeveloped. As Carnot would complain in 1824, for conventional machines there existed "a mechanical theory [that] permits a very detailed study.... Every situation can be foreseen, and every conceivable movement is subject to some well-founded and universally applicable general principle. It is this which characterizes a complete theory, but such a theory is plainly lacking in the case of the heat engine" (Carnot, 1986 [1824], p. 64). There were nonetheless a number of strands of activity in the late eighteenth and early nineteenth centuries that would contribute to such a "complete theory." Developments within both practical and theoretical engineering stimulated both the refinement of a concept of mechanical "work" and the analysis of the determinants of the "efficiency" of various types of machine. British attempts to understand the new "expansive" mode of operation of high-pressure steam engines were very suggestive, and Carnot's work emerged directly from a similar French debate.

Concepts of Energy and Work

The concept of a generalized physical "energy" was only constructed in the mid-nineteenth century, but even the more-limited concept of mechanical "work"—the product of a force and the distance through which it acts—was only gradually articulated during the eighteenth and early nineteenth centuries.

There was no clear concept of "energy" in Newton's mechanics. For Newton, the conserved quantity of motion was the *momentum* of a body, calculated as the product of its mass and velocity. On the other hand, his great rival, the German philosopher and mathematician Gottfried Wilhelm Leibniz (1646–1716), proposed that a better measure of the quantity of motion was *vis viva* ("living force"), the product of mass and the *square* of speed; this is similar, of course, to the modern concept of kinetic energy ($1/2 \, mv^2$). Vigorous and productive, but ultimately unresolved, controversy ensued through the early eighteenth century. An obvious objection to Leibniz's claims was that his *vis viva,* in contrast to momentum, was only conserved in perfectly elastic collisions. Leibniz could only respond that the *vis viva* that seemed to disappear was in fact distributed through the microscopic parts of a body. Such issues played an important role in late-eighteenth-century discussions of efficiency.

Similarly, the concept of "work" had little clear role in the academic "rational mechanics" so vigorously developed in the aftermath of Newton's *Principia* by the likes of Leonhard Euler (1707–83) and Joseph-Louis Lagrange (1736–1813). Instead the concept gradually emerged mainly from the practical activities of engineers and was given increasingly precise expression by writers of textbooks on engineering mechanics, especially in France, between the 1780s and 1820s.

For British engineers like Watt, for example, an obvious expression of the power or "duty" of an engine—initially a pump more often than not—was the weight of water in pounds (lb.) that could be raised in a given time, multiplied by the distance in feet (ft.) through which it was lifted; hence a common measure was in terms of foot-pounds per minute. More vaguely, the capabilities of a machine could be expressed in terms of the man or animal power that it could replace, that is, in terms of its "manpower" or "horsepower." Initially neither term was more precisely defined, and there was not even agreement on the ratio of man to horse; in Britain one horsepower was deemed equivalent to the power of five men, in France to seven men—although it is not clear whether this reflects well on British men or on French horses. An entrepreneurial engineer like Watt had an interest in defining the "power of a horse" more precisely, in terms of its ability to raise a given weight through a given height in a given time. Although Watt was not the first to do this, his figure of 33,000 foot-pounds per minute as the "duty" to be expected from a horse was widely accepted.

From French engineers, more professional and more systematically trained in academic terms, there flowed from the 1780s onwards a stream of textbooks on engineering mechanics. It was mainly within this tradition that there emerged both a clearly defined general concept of "work" in the modern sense and also new metric units for its measurement.

The tradition could be said to have begun with the 1783 *Essay on machines in general* by the French engineer and radical politician Lazare Carnot (1753–1823), known as "the Organizer of Victory" for his work on behalf of the young Republic in the 1790s, and the father of Sadi Carnot. In a narrow terminological sense the tradition could be said to culminate in Gaspard Coriolis' (1792–1843) *On the calculation of the effect of machines* of 1829, which explicitly employs the term "work" (*travail*) to mean "the quantity which is fairly commonly called mechanical power, quantity of action or dynamical effect." This "work" was measured as the product of a force (still usually a weight) and the distance (still usually the height) through which it acted. For J.N.P. Hachette (1769–1834), in his 1811 *Elementary treatise on machines*, a convenient (for an engineer) unit of work was the "dynamode," equal to the effort required to raise a weight of 1,000 kilogram through 1 meter (and thus equal to 9.8 kJ). This concern with the precise, standardized quantification of physical variables reflected the same preoccupation with the commercial and industrial function of science that motivated the Revolutionary "metric" system.

The French engineering tradition emphasized the correlation between the work done by a descending weight and the *vis viva* that could potentially result. But now it was the *work* that was taken as fundamental. *Vis viva* was increasingly defined as $1/2 \ mv^2$ (instead of as just mv^2); according to Coriolis, "If they [mathematicians] previously used the name of 'living force' for the product of mass and velocity squared, it is because they did not pay attention to *work*. . . . All practitioners today mean by 'living force' the work which can produce the velocity acquired by a body. . . ." (Cardwell, 1966, p. 219). Change of height (h) and speed (v) were related by the equation: $mgh = 1/2 \ mv^2$, instead of the equation $2mgh = mv^2$ usually found in the eighteenth century.

The established conservation of *vis viva* could thus be broadened into a more general conservation of "mechanical effect." In his 1807 *Course of Lectures on Natural Philosophy and the Mechanical Arts*, Thomas Young introduced the everyday word "energy" into mechanics, labeling *vis viva* as "actual energy," correlated with the "potential energy" that a body possessed by virtue of its height. But this remained a quite narrow usage. As Thomson later pointed out: "The very name 'energy,' though first used in its present sense by Dr. Thomas Young about the beginning of this century, has only come into use practically after the doctrine which defines it had . . . been raised from a mere principle of mathematical dynamics to the position it now holds of a principle pervading all nature and guiding the investigator in every field of science" (Thomson, 1881, p. 513; Smith, 1998, p. 8).

Concepts of Efficiency

Theoretical discussion of the "efficiency" of machines—that is, the relationship between the "effort" (or input) and the "effect" (or output)—also developed during the eighteenth century. Efficiency also was primarily the concern of engineers rather than academic mathematicians, although there was some overlap with the issue of the conservation or dissipation of *vis viva* in collisions. Methods of estimating efficiency for water power and for steam engines were inevitably different.

In the case of steam engines, before the acceptance of a "mechanical equivalent of heat" (see chapter 4), it was not possible directly to compare thermal input and mechanical output. Efficiency could only be expressed in terms of the greater or lesser amount of work done for a given amount of fuel. Thus, as already noted, the performance of engines was commonly expressed in terms of foot-pounds of work done for every "bushel" (a widespread, but by no means uniform, measure of weight or volume, generally about 8 gallons or 84 pounds) of coal burned. In trials in 1772 and 1778, for example, Smeaton measured the duty of a traditional atmospheric engine as 5,044,158 foot-pounds per bushel, the duty of his own improved replacement engine as 9,636,660, and that of Watt's new engine with separate condenser as 18,902,136.

For steam, theoretical calculation of the potential output awaited the nineteenth century. In the case of water power, however, it was possible to go beyond measurements in field trials. A water-wheel could be regarded as a kind of mechanical device like a pulley or winch, for which there was a long tradition of calculating the expected relationship between input and output. As early as 1704 Antoine Parent (1666–1716) had argued—erroneously but very influentially—that on theoretical mechanical grounds a maximum "effect" of only 4/27 of the "effort" was achievable for any water-wheel. In 1752, however, a French engineer, the Chevalier Antoine Deparcieux (1703–68)—while working on the water supply at the chateau of the king's mistress, Madame de Pompadour—reasoned, by analogy with two weights connected over a simple pulley, that there was no fundamental reason why any water-wheel should not be able to drive an identical wheel of buckets raising water back up to the mill pond; a certain weight of water descending through a water-wheel could ideally raise exactly the same weight back up to the original level again. It should therefore be possible to recover *all* the "effort" of the water and approach an efficiency of one.

It was equally clear, however, that this limit could never be achieved in practice. In a neat illustration of the contrasting approaches of French and English engineers, Smeaton in 1759 set about investigating the performance of different types of water-wheel by experiments with scale models and the painstaking systematic variation of rates of flow and other variables. There were three main types of water-wheel in use: the over-shot, the under-shot, and the breast wheel. Smeaton demonstrated that the overshot wheel was the most efficient, the undershot the least. More generally, he concluded that in order to maximize efficiency it was essential to minimize impact, turbulence, and the relative motion of wheel and water. Any residual motion of the water leaving the wheel represented unused effort. Smeaton's studies resulted in the general adoption of the over-shot wheel where possible.

Similar ideas emerged within a more theoretical and strictly mechanical framework in Lazare Carnot's *Essay*. Carnot's approach to a *practical* understanding of the operation of machines was based on the conservation or dissipation of *vis viva*. Reflecting on the dissipation of "living force" in collisions, he emphasized the importance of eliminating percussion or turbulence in order to achieve maximum efficiency. The main academic precedent for this perception had been the science of hydrodynamics developed from Daniel

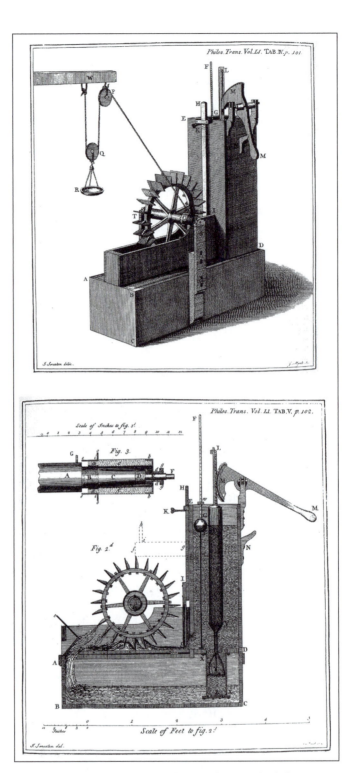

Figure 3.1: John Smeaton's scale model with which he measured the efficiency of an undershot waterwheel, from his 1759 paper "An experimental Enquiry concerning the natural Powers of Water and Wind to turn Mills, and other Machines." *Philosophical Transactions of the Royal Society*, vol. 51; by permission of the Syndics of Cambridge University Library.

Bernoulli's *Hydrodynamics* of 1738. Thus, in the work of Deparcieux, Smeaton, and Carnot Senior one can see the origin of ideas that would be crucial to Sadi Carnot's treatment of the steam engine. There is the notion of two identical machines operating in reverse, and the insistence on the importance of infinitesimally imbalanced—in other words, reversible—operation to minimize losses. It is intriguing that the development of thermodynamics should owe so much to the water-wheel as well as to the steam engine.

British Speculation: Watt, Robison, and "Expansive" Operation

Alongside a constant pragmatic striving for increased efficiency, efforts to reach a more analytical understanding of the operation of the steam engine seem to have originated in England towards the end of the eighteenth century. In 1797 Joseph Black's colleague and editor John Robison contributed a long article on "the Steam Engine" to the third edition of *The Encyclopaedia Britannica.* Apart from much descriptive historical and technical detail, his article includes a precise and elegant analysis of the work done and the superior economy to be obtained by the "expansive" operation of a steam engine. Robison took it for granted that the output of an engine—which he also denoted in general terms as "work"—could be calculated as the "accumulated pressure," that is, as the product of the pressure within the cylinder and the volume swept by the piston in the active stroke. He took a diagrammatic cross-section of a cylinder and used it as a graph of the pressure of the steam and of the volume swept by the piston. Like Watt, Robison was very interested in the possibility of fuel economy by the "expansive" operation of an engine, that is,

Figure 3.2: James Watt's diagram of the "expansive" operation of a steam piston for his 1782 patent application; the steam supply was to be cut off early at K, after only one quarter of the full stroke of the piston. The cross-sectional diagram of the piston served as a graph or "curve, the [horizontal] ordinates of which represent the powers [pressures] of the steam when the piston is at their respective places;" the area under this curve represented the power delivered by each stroke of the piston. James P. Muirhead *Mechanical Inventions of James Watt,* vol. 3 (London, 1854); by permission of the Syndics of Cambridge University Library.

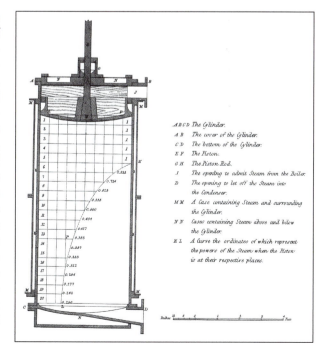

by cutting off the steam supply to the cylinder at an early stage in the motion of the piston—after 1/4 of the power stroke, say—and then allowing the steam to expand, its pressure gradually decreasing as it did so. With some elegant mathematics, Robison was able to show that operating in this fashion "one-fourth of the steam performs nearly 3/5ths of the work, and an equal quantity [of steam] performs more than twice as much work when thus admitted during 1/4th of the motion" (Robison, 1797, p. 765).

Robison's analysis was closely connected to actual engineering practice. Since the 1790s in Britain gauges had been fitted to cylinders in order to monitor the changing pressure within them, and thereby to estimate the power that might be produced. One such gauge comprised a small spring-loaded piston with a pointer attached which moved against a scale to indicate the pressure. In the early 1780s Watt wrote to Boulton: "If you have a notion that young Southern would be sufficiently sedate, would come for a reasonable sum annually, and would engage for a sufficient time, I would be very glad to engage him... provided he gives bond to give up music, otherwise he will do

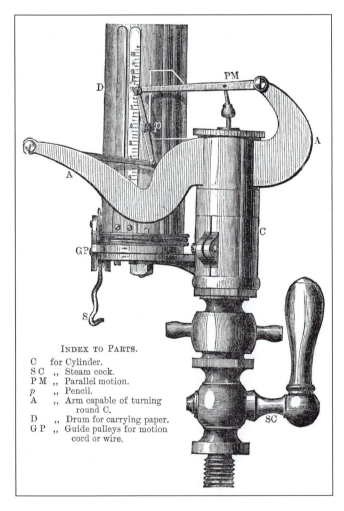

INDEX TO PARTS.

C   for Cylinder.
S C   ,,   Steam cock.
P M   ,,   Parallel motion.
p   ,,   Pencil.
A   ,,   Arm capable of turning round C.
D   ,,   Drum for carrying paper.
G P   ,,   Guide pulleys for motion cord or wire.

Figure 3.3: A "steam indicator" for monitoring the output of a working steam engine; the curve drawn by the recording pencil (p) is just visible on the upper vertical cylinder; compare with figures 3.2 and 3.5. Andrew Jamieson (London, 1900) *Elementary Manual on Steam and the Steam Engine*; by permission of the Syndics of Cambridge University Library.

no good, it being a source of idleness" (Smith, 1998, p. 33). Presumably John Southern (1758–1815) must have given the required undertaking, for he became Watt's assistant and in 1796 invented a device, subsequently known as the "steam indicator," in which a pencil replaced the pointer and thus traced a line recording the pressure on a sheet of paper that was moved backwards and forwards by the piston. In addition to providing a "print-out" of the variations in pressure with the motion of the piston, the area within the curve drawn by the indicator pencil represented the work done by each stroke of the piston, just as Robison's diagram had done. Such indicators proved extremely useful in measuring the performance of an engine (without the trouble of extensive field trials) and, when setting up a new engine, allowed the fine-tuning of valve settings to maximize performance. Not surprisingly, therefore, Watt strove to keep the device a secret, and with considerable success: The first printed account did not appear until 1822.

French Speculations after 1815

When the high-pressure steam engine was introduced to France after the fall of Napoleon in 1815, it stimulated renewed discussion of the theory of steam engines in general, and of the superior efficiency of the high-pressure engine in particular. Several of the elite academic scientists whose work on heat and gases has been described in chapter 2 tackled the issue. One of the first was Petit in 1818; he calculated the work that had to be done in pushing back the atmosphere when 1 cm^3 of water boiled into 1700 cm^3 of steam; this, he suggested, represented the maximum amount of work that could possibly be extracted from a given volume of steam. Unfortunately, when combined with accepted figures for the amount of water that a given amount of coal could boil, the amount that he calculated suggested a "duty" that was less than that already achieved by some of the best imported engines of the day.

Petit's conclusions were criticized by Clément and Desormes on the grounds that he had neglected the contribution of expansive operation, which was perceived to be crucial to the operation of high-pressure engines. In their alternative treatment in 1819 they imagined a bubble of steam rising through a tank of water and expanding as the pressure decreased; the work done was to be measured by the amount of water displaced by the steam bubble. The published account of their work considers the expansion of the steam to be *isothermal*—as did Robison and most other writers until the 1840s. However, it appears that in unpublished lectures that his friend Sadi Carnot may have attended, Clément regarded the expansion as *adiabatic, with consequent cooling of the steam*. These treatments of the problem, and others at the time, all overlooked the waste of heat from the hot steam exhausted.

There was also much lively speculation in many quarters, both academic and commercial, about the possibilities of using working substances other than steam in a heat engine. Air was perhaps the most plausible alternative, as employed in the "Stirling" engine patented in 1816 by the Presbyterian minister the Rev. Robert Stirling (1790–1878), and considerable attention was also directed to the apparently "irresistible" force of solids expanding when heated, or of ice when freezing. (One's attention is particularly caught by proposals for the use of alcohol as the working substance [not as fuel], which at first sight seems as promising as the combined hot-air and hydrogen balloons trialed in the late eighteenth century and would have required very careful selection of operatives, even in a Presbyterian community.)

Sadi Carnot moved on the periphery of the domain of elite French scientists, and his masterpiece emerged from the context of the above discussions.

SADI CARNOT'S *REFLEXIONS ON THE MOTIVE POWER OF FIRE*

Nothing in the whole range of Natural Philosophy is more remarkable than the establishment of general laws by such a process of reasoning. (Thomson, 1882–1911, 1: pp. 113–4 [1849], commenting on Carnot's theories; Smith, 1998, p. 88)

Sadi Carnot's *Reflexions on the motive power of* fire of 1824, as mediated in the first instance by Clapeyron's "Memoir" of 1834, provided a crucial stimulus to the subsequent foundation of the modern science of thermodynamics from the late 1840s onwards. In the most general terms, Carnot's work suggested a new kind of science of heat. Rejecting that focus on the microscopic structure of matter, caloric atmospheres, and so on that was typical of the Laplacian tradition, Carnot dealt rather with such directly measurable macroscopic quantities as temperature, pressure, heat, and "motive power." In more particular terms, he articulated a number of insights and techniques that were to become essential components of later thermodynamics—the crucial importance of temperature *difference* to the operation of heat engines, for example, and the idea of reversible operation and the closed "Carnot cycle." From these ideas Carnot himself arrived at a range of interesting conclusions—especially that the working substance was irrelevant to the maximum efficiency of a heat engine.

Carnot and Clapeyron: Professional Engineers

Sadi Carnot and Clapeyron both followed careers typical of the new French breed of professional engineer—although in conventional terms Clapeyron was much the more successful. In his own time Carnot would have been regarded as a very minor figure, very much on the fringe of the elite Parisian scientific community based around the Institute and the École Polytechnique. Carnot was the eldest son of the "Organizer of Victory" Lazare Carnot. After studying at the École Polytechnique and at the military School of Artillery and Engineering, Carnot Junior joined the army as a military engineer. Always reserved, it is doubtful that he found military life congenial, and in 1819 he moved to Paris on permanent leave of absence. Here he was able to pursue his scientific studies, attending public lectures and making friends with Nicolas Clément.

Carnot died prematurely, probably of cholera, in 1832. It is possible that he had met Clapeyron in that year. Like Carnot, Clapeyron studied at the École Polytechnique before progressing to another state engineering school, the School of Mines. He spent most of the rest of his working life as a railway engineer, with a special interest in the design of steam locomotives, and he was elected a member of the Academy in 1848. Apart from the reworking of Carnot's *Reflexions* in his "Memoir on the motive power of heat" of 1834, which appeared in English translation in 1837, Clapeyron is best known for work on vapor pressure.

Carnot became particularly interested in issues of industrial development and, from 1821, in the workings of the steam engine, whose "importance is immense, and their use is increasing daily. They seem destined to bring about a great revolution in the civilized world" (Carnot, 1986 [1824], p. 61). Not surprisingly, given the recent Napoleonic wars, his interest was colored by economic and military rivalry with Britain: "If you were now to deprive England of her steam engines, you would deprive her of both coal and iron; you would cut off the sources of all her wealth, totally destroy her means of prosperity, and reduce this nation of huge power to insignificance. The destruction of her navy... would probably be less disastrous" (ibid., p. 62). But he also inherited his father's fraternal Enlightenment optimism. "There is a sense," he suggested, "in which steam navigation brings the most widely separated nations closer together. It tends to unite the people of the world and make them dwellers, as it were, in one country" (ibid., p. 62).

His meditations upon the steam engine resulted in 1824 in his only published work, whose full title is *Reflexions upon the motive power of fire and upon machines designed to develop that power.* As already recorded, Carnot's ultimate aim was to construct "a complete theory" of heat engines. Despite the rapid and already highly sophisticated development of the heat engine, he claimed, "the theory of its operation is rudimentary, and attempts to improve its performance are still made in an almost haphazard way" (ibid., p. 63). A different approach was needed. "In order to grasp in a completely general way the principle governing the production of motion by heat, it is necessary to consider the problem independently of any mechanism or any particular working substance. Arguments have to be established that apply not only to steam engines but also to any conceivable heat engine, whatever working substance is used and whatever operations this working substance is made to perform" (ibid., p. 64).

Despite the generality of his ambitions, Carnot was also passionately concerned to be of use to the working engineer, and he identified two particularly important practical issues. Firstly: "The question whether the motive power of heat is limited or whether it is boundless has been frequently discussed. Can we set a limit to the improvement of the heat engine, a limit which, by the very nature of things, cannot be in any way surpassed? Or, conversely, is it possible for the process of improvement to go on indefinitely?" And secondly: "For a long time there have also been attempts to discover whether there might be working substances preferable to steam for the development of the motive power of fire...Might air, for example, have great advantages in this respect?" (ibid., p. 63). These questions, as noted above, were current in both practical and academic circles in Carnot's time. In answering them, he drew upon both the academic debate—especially the ideas of Clément—and upon the older tradition of engineering mechanics that his father Lazare had to a large extent initiated. But Sadi's synthesis of the various strands was very distinctively his own.

Heat Engines and Water-Wheels: The Logic of the *Reflexions*

The beauty of Carnot's slim book is that profound and persuasive conclusions are drawn from assumptions that appear at first sight unexceptional, even trite and unpromising. One is reminded of that most French of philosophers, Descartes, who insisted that a tenacious search for "clear and distinct ideas" would inevitably lead to the truth. This precise scrutiny of the operation of the heat engine led Carnot to a range of profound conclusions, specifically, the importance of temperature difference, the importance for optimum efficiency of reversible operation, and thence that there must be a fixed maximum efficiency for all heat engines, dependent solely on the temperatures at which they operate and independent of their particular design or working substance.

It is ironic, given the radical consequences, that the model that underpins much of Carnot's highly original speculation will appear doubly archaic to modern eyes. Firstly, Carnot accepted the *caloric theory of heat:* Heat was to be regarded as a fluid, and above all a fluid that is conserved. His acceptance of this theory—although by no means uncritical—will become quite evident in the ensuing account. Secondly, it was the operation of *water-powered* engines, such as the water-wheel, that informed much of his thinking on the heat engine. Thus, towards the end of his initial analysis of the nature of the heat engine, Carnot concluded that "we are sufficiently justified in comparing the motive power of heat with that of a fall of water" (ibid., p. 72), and hydrodynamic arguments and analogies abound throughout.

Carnot's first major insight was that it was not just heat as such but a *difference* in temperature that was essential to the working of the steam engine:

> The production of motion in the steam engine always occurs... when caloric passes from a body at one temperature to another body at a lower temperature. So what exactly happens in a steam engine of the kind now in use?.... the production of motive power in a steam engine is due not to an actual consumption of caloric *but to its passage from a hot body to a cold one.* ... It follows from this principle that, in order to create motive power, it is not enough simply to produce heat. Cold is also essential; without it, the heat is useless. (ibid., pp. 64–65; original italics)

This is perhaps less obvious than it seems. In the early "fire engines" it must have seemed to be precisely the hot *fire* as such that did the work. Equally astute was Carnot's further perception that the transfer of heat from a higher to a lower temperature *without work being done* automatically entailed a *waste* of possible "mechanical effect."

> Wherever there is a difference in temperature, motive power can be produced.... Since any process in which the equilibrium of caloric is restored can be made to yield motive power, a process in which the equilibrium is restored without producing power must be regarded as representing a real loss.... [T]he direct transfer of caloric from a hotter body to one that is colder... occurs most commonly when bodies at different temperatures are in contact with one another, and contact of this kind must therefore be avoided, so far as possible. (ibid., pp. 67, 70)

In the operation of any heat engine, therefore, temperature differences should be kept infinitesimal (or minimal); the engine should operate through a series of almost-at-equilibrium or "quasi-static" states. In a condenser, for example, the condensing reservoir and the steam to be condensed should be at the same temperature, for "the most minute difference in temperature is enough to bring about condensation" (ibid., p. 68n).

This condition for minimizing waste in the operation of a heat engine was very similar to the conditions proposed by Smeaton and Lazare Carnot for the most efficient operation of water-wheels (and machinery in general): Ideally there should be minimal difference in velocity between the water and the wheel; there should be no impact or turbulence. But did such a procedure necessarily guarantee the maximum possible output of useful work from a given quantity of caloric? And, even if relevant to the operation of any given type of heat engine, did it cast light upon the maximum possible efficiency of different types of heat engines in general? Carnot believed that it did.

He proceeded to argue that, if the condition of infinitesimal temperature difference were met, then it would automatically be possible to *reverse* the operation of any such heat engine *exactly*. It was well known that this would apply in principle to a frictionless, mechanical engine. According to Carnot, the same could be true for a heat engine. Not only does a temperature difference enable motive power to be produced, he argued, but

> [t]he converse is also true: wherever there is power which can be expended, it is possible to bring about a difference in temperature and to disturb the equilibrium of caloric... It is an experimental fact that the temperature of gaseous substances rises when they are compressed and that it falls when they are expanded. In this way, changes in temperature can be brought about and the equilibrium of caloric can be disturbed as often as we like, using the same substance. (ibid., p. 67)

Consequently, Carnot claimed, because the operation is virtually at equilibrium, the slightest variation in temperature, or in the input or output of power, will push the process in one direction or the other. Since it was still not universally agreed that adiabatic heating was independent of the *speed* of operation, this was quite a bold generalization.

On this basis Carnot conceived an ideal steam engine, performing mechanical work as caloric was transported from a hot reservoir A (the "furnace") to a cold reservoir B (the "refrigerator") in three stages: isothermal formation of steam at the temperature of the furnace, adiabatic expansion of the steam (with a resultant cooling to the temperature of the refrigerator), and isothermal condensation of the steam into water. Always assuming the absence of any mechanical wastage through friction and so on, such an engine could then operate equally well in reverse, using the same amount of mechanical effort in compression as had been released in the previous expansion to transport the same amount of caloric back from the cold to the hot reservoir. Thus, two such engines operating alternately or in tandem would cancel each other out. "It would be possible to perform any number of these operations alternately without, in the end, producing motive power or transferring caloric from one body to the other" (ibid., p. 69).

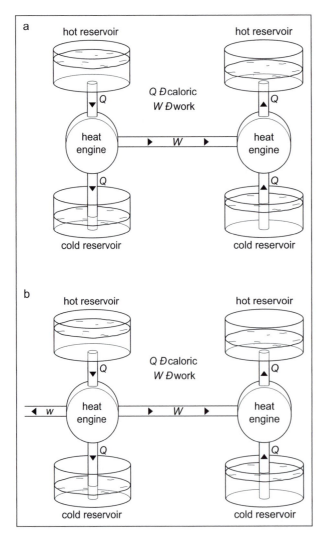

Figure 3.4: Sadi Carnot's argument that a reversible engine is the most efficient that there can possibly be:

a. Two reversible heat engines working in opposition (between the same temperatures) cancel each other out exactly. (NB. For Carnot, the quantity of heat or caloric that flows through each engine is conserved: the heat is not used up in generating work, nor vice versa.)

b. A "super-efficient" engine on the left combined with a reversible engine generates a surplus of work, whilst the caloric simply circulates between the hot and cold reservoirs. Illustration by Jeff Dixon.

This scenario clearly established, Carnot was able to return to his starting questions: Was there an intrinsic limit to the possible improvement of the heat engine? And did this depend upon the type of engine and the working substance? His answers were: Yes, we can set a limit; and no, it does not matter what working substance we use.

He argued that the reversible engine just described must produce the maximum possible "effect" for a given quantity of caloric operating between two given temperatures.

> If there existed any better method than this for utilizing heat…then we should merely have to abstract some of this power in order to return the caloric from the body B back to the body A, using the method just outlined. In this way, the caloric would be made to pass from the refrigerator to the furnace, and the initial conditions would be restored. If we did that, we should then be in a position to begin again the identical process all over again, and to continue in the same way indefinitely. We should then have a case not only of perpetual motion but of motive power being created in unlimited quantities without the consumption of caloric or of any other agent. Creation of this kind completely contradicts prevailing ideas, the laws of mechanics, and sound physics; it is inadmissible. (ibid., p. 69)

The reversibility of the quasi-static engine, therefore, guaranteed that it was the most efficient possible engine. Carnot concluded, therefore, in answer to his first main question, that there was an intrinsic limit to the efficiency of the heat engine, "that caloric cannot be made to yield a greater quantity of motive power than is obtained by the…sequence of operations we have just described."

And the same argument could be equally applied not just to a pair of engines of the same type, but to any two engines, and would thus answer his second main question. No combination of two engines using different working substances could possibly generate surplus work; they must all have the same maximum efficiency, the same limit must apply whatever the design or working material used. In other words, "the following general proposition may be stated: *The motive power of heat is independent of the working substances that are used to develop it. The quantity is determined exclusively by the temperatures of the bodies between which, at the end of the process, the passage of caloric has taken place*" (ibid., pp.76–77, original italics). This conclusion must have seemed surprising and unhelpful, at least to engineers who were well aware of the big practical differences between the construction and operation of engines using air instead of steam.

The Carnot Cycle and Ideal Gases

However unexpected the above conclusion, it allowed Carnot to further extend his science of the motive power of fire. If all substances would perform the same in an ideal heat engine, then Carnot could choose for further investigation a substance whose behavior was already well understood, namely, the ideal gas that had been the object of intense scrutiny during the first two decades of the nineteenth century (see chapter 2).

So Carnot reformulated the arguments that had been based on an incomplete three-stage sequence of steam processes in terms of a complete *closed* four-stage cycle gas engine. He was partly motivated by his awareness of "a more substantial objection to our proof" (ibid., p. 71), namely, that the process he described for the steam engine was incomplete: At the end of each stroke, "if we wish to start a similar process all over again to develop more motive power with the same instrument, the same steam, we must begin by restoring the original conditions; the temperature of the water must be made once again what it was at the start" (ibid., p. 71). This could be achieved by returning the condensed water to the furnace, but this would be an *irreversible* process. Without a complete, closed cycle, it was impossible to conduct a wholly satisfactory comparison of the mechanical power developed for a given amount of heat.

To avoid these criticisms, he described a theoretical heat engine that operated in a closed loop with ideal gas as the working substance, *every* stage of the operation being reversible. This was the now familiar "Carnot cycle." Once established, the cycle consisted of four stages: isothermal expansion (from C to E in figure 3.5), adiabatic expansion (from E to F), isothermal compression (from F to K), and finally adiabatic compression (from K back to C). The first two stages—isothermal followed by adiabatic expansion—clearly reflected the newly widespread practice of "expansive" operation. The obvious antecedent for treating the second "expansion" as *adiabatic* was the work of Carnot's friend Clément. The attraction for Carnot was that it enabled a *reversible* transition from one temperature to another.

Apart from a diagram of a piston, Carnot's exposition was very largely verbal; the pressure-volume graph that is now commonly associated with Carnot's cycle—the area within which represents the work done—was only introduced by Clapeyron (see figure3.5). His graph is in essence a neat, mathematical

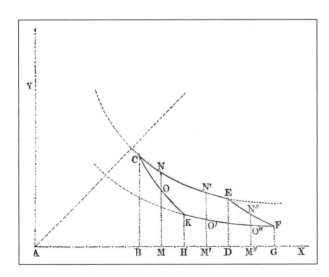

Figure 3.5: Émile Clapeyron's graph of the Carnot cycle, in which Carnot had applied his general principles to the cyclical operation of an ideal-gas engine. Clapeyron (1837 [1834]), "Memoir on the Motive Power of Heat," in R. Taylor, ed., *Scientific Memoirs*, vol. 1; by permission of the Syndics of Cambridge University Library.

version of the diagram produced by steam indicator; although the precise debt is uncertain, Watt's device and its associated diagram, a little known trade secret in Carnot's day, had become quite widely known ten years later when Clapeyron was writing.

Having described his cycle, Carnot's next step was to restate his earlier, possibly flawed, arguments for the maximum efficiency of the reversible engine, and for the consequent irrelevance of the working substance. Using these principles and his four-stage cycle, he moved on to an extended discussion of the thermal behavior of ideal gases. He derived a series of conclusions, most of which have their equivalents in modern thermodynamics. A couple of his conclusions will be sketched briefly in order to give some indication of his approach.

Firstly, Carnot claimed that: "When a gas passes from a particular volume and pressure to another specified volume and pressure, without undergoing a change in temperature, the amount of caloric absorbed or given out is always the same, *whatever the gas on which the observation is made*" (ibid., p. 78). This led fairly directly to a second proposition: "The difference between the specific heat at constant pressure and the specific heat at constant volume is the same for all gases" (ibid., p. 80). In modern texts this would be expressed as (c_p-c_v) is constant.

Both of these conclusions would be valid in modern thermodynamics based on the conservation of energy and the equivalence of heat and mechanical work. Such a foundation was not available to Carnot—indeed, he would have categorically rejected its validity. His argument, therefore, was based rather on his own new *efficiency* principles. "Since two different [ideal] gases at the same temperature and pressure must behave identically in the same conditions, it follows that if both of them undergo the processes just described, they must each yield the same quantity of motive power. This implies, in accordance with the fundamental propositions we have established, that equal quantities of caloric must be used in the two cases" (ibid., p. 78). In other words, any two different ideal gases, since they both obey the same equation of state, must do the same work if they move between the same two states; to be equally *efficient*—as Carnot's new principles demanded—they must therefore absorb or release the same amount of heat.

Four years later, in 1828, and apparently quite independently, Dulong tentatively arrived at similar conclusions on the basis of an indirect reinterpretation of his own new measurements of the ratio of specific heats, γ and established data on c_p. Strangely (in retrospect, at least), neither Carnot nor Dulong nor any of their immediate contemporaries remarked upon the possible implications of the equality of the work done and the heat absorbed. However, the independent experimental confirmation of Carnot's theoretical conclusions did make a big impression on Clapeyron and later thinkers (see chapters 4 and 5).

Finally, Carnot returned to the question of how the motive power developed by the fall of caloric depended on the temperature difference between the hot furnace and the cold refrigerator. It seemed clear that the quantity of work developed would be greater for a greater temperature difference, but would it be the same for the same difference at higher or lower temperatures? Would a fall from 100°C to 50°C produce the same "effect" as a fall from 50°C to 0°C? The analogy of the water-wheel would suggest that the output would be the same in each case, but Carnot suspected otherwise. On the basis of a complex argument, which depended crucially on Delaroche and Bérard's conclusion that the specific heat (by weight) of a gas decreases as its density increases, Carnot concluded that *"the fall of caloric yields more motive power at lower temperatures than it does at higher ones"* (ibid., p. 93; original italics). Despite the flawed premises, whether by luck, intuition, or practical experience, this conclusion is in fact valid in modern thermodynamics. In a rare mathematical footnote Carnot attempted to derive an analytical expression for the efficiency, e. As reworked by Clapeyron, this expression became $e = \Delta\theta/C(\theta)$, where $\Delta\theta$ was the temperature difference and $C(\theta)$ was an as yet undetermined "Carnot function" of the higher temperature. The search to identify the exact nature of this function—which eventually turned out to be equal simply to the absolute temperature, T—became a vital part of the early construction of thermodynamics (see chapter 5).

THE RECEPTION OF CARNOT'S THEORY

I went to every bookshop I could think of, asking for the *Puissance Motrice du Feu*, by Carnot. "Caino? Je ne connais pas cet auteur." ... the *Puissance Motrice du Feu* was quite unknown. (William Thomson, 1882–1911, 2: p. 458n [1892], on his search for a copy of Carnot's *Reflexions;* Smith, 1998, p. 44)

Carnot, Clapeyron, and Regnault

Carnot's *Reflexions* received a couple of favorable reviews upon publication, but was then very largely ignored until Clapeyron's reworking a decade later, which was initially ignored in its turn, although it was accessibly published in English in 1837. It is not entirely obvious why Carnot's original and powerful methods and his sometimes striking conclusions should have been so completely ignored. It may be that his approach fell between two stools, between academic physics on the one hand and practical engineering on the other. His arguments were very largely presented in verbal or numerical form. This limitation was remedied by Clapeyron, who presented Carnot's arguments in the calculus terms more familiar (and respectable) in the world of academic physics, and, as already noted above, he also introduced the now familiar "steam indicator" diagram of the Carnot cycle. Moreover, from a study of the latent heats of vaporization of various liquids Clapeyron was able to estimate values for the Carnot function, $C(\theta)$, at different temperatures; these values seemed to confirm Carnot's conclusion that efficiency would decrease at higher temperatures.

Even Carnot himself seems to have been reluctant to promote his ideas, possibly because of his growing doubts about the validity of the caloric theory that was fundamental to his speculations. In manuscripts notes only published after his death, Carnot wrote: "Heat is nothing but motive power, or rather motion, which has changed its form. It is motion of the particles of bodies. Wherever motive power is destroyed, there is a simultaneous production of an amount of heat exactly proportional to the motive power that is destroyed. Conversely, wherever there is destruction of heat, motive power is produced" (Carnot, 1986, pp. 26, 191). For whatever reason, Clapeyron too took little further interest in developing the topic after the "Memoir," even in his own lectures. Thus, it was not until the late 1840s that Carnot's ideas were rescued from oblivion, principally by the young William Thomson.

The spirit of French scientific study of heat in the following generation is embodied in the person of Henri Victor Regnault (1810–78). Initially an organic chemist, a passing interest in the specific heats of gases was crystallized into his life's work by "the fatal step of accepting financial support for his experiments from the Ministry of Public Works. Encouraged by the generous official backing, he immediately embarked on the systematic re-determination of all the experimental data that could conceivably be required in the theory and practice of steam-engines and other heat engines. The result, as it emerged over the next twenty-five years, was the most precise and yet the most unimaginative of compilations, a warning perhaps of the dangers of dependence on government sponsorship in the pursuit of science" (Fox, 1971, p. 298). Pursuing the precise, quantitative, experimental, antispeculative trend of Dulong in the 1820s to the extreme, Regnault's approach was conspicuously untheoretical and thus, to some perhaps, unglamorous. Nevertheless, according to Fox (1971, p. 295), "when Regnault ceased experimental work about 1870, he had provided answers to nearly all the most important [empirical] problems relating to the study of heat which had been tackled, with such limited success, since the middle of the eighteenth century. That his answers have undergone only minor modification to this day bears witness to the excellence of his work." Regnault's almost industrial approach to research, although in a tradition initiated in the Napoleonic era, thus set new standards of thoroughness and precision in experimental physics. His data was an essential foundation for the subsequent development of thermodynamics. Many of the next generation of physicists, including the recently graduated Thomson, passed through his laboratory.

Thomson and the Absolute Temperature Scale

It was in 1845, after a brilliant conclusion to his undergraduate mathematical studies at Cambridge, that William Thomson visited Paris to gain practical laboratory experience under the master experimentalist Regnault. At some point he encountered Clapeyron's "Memoir," although, as we have heard above, he was at first unable to find a copy of Carnot's original book. Thomson realized that their ideas would allow the solution of one important problem in the science of heat, the problem of establishing an absolute scale of temperature.

All temperature scales up to this time had been more or less arbitrary, based on the behavior of some specific substance. But neither the expansion of mercury, nor of alcohol, nor of air or any other gas could claim to register changes in temperature in any absolute sense. Even the very similar behaviors of the "permanent" gases, from which the concept of the "ideal" gas was derived, offered an only slightly more secure foundation. As Thomson saw, however, Carnot's conclusion that the theoretical efficiency of any heat engine was totally independent of its working substance provided the basis for an entirely general temperature scale, independent of the properties of any specific substance. He explained his idea in an 1848 paper, "On an absolute thermometric scale, founded on Carnot's theory of the motive power of heat, and calculated from the results of Regnault's experiments on the pressure and latent heat of steam."

> The characteristic property of the scale which I now propose is, that all degrees have the same value; that is, that a unit of heat descending from a body A at the temperature $T°$ of this scale, to a body B at the temperature $(T-1)°$, would give out the same mechanical effect [work], whatever be the number T. This may justly be termed an absolute scale, since its characteristic is quite independent of the physical properties of any specific substance. (Thomson, 1882–1911, 1: p. 104 [1848]; Smith, 1998, pp. 51–2)

In other words, a degree of temperature anywhere on the scale would be defined as the temperature difference that would result in a fixed amount of work being done by a unit of heat falling through it. Although as originally conceived by Thomson the new absolute scale did not tally with the ideal gas temperature scale, the two scales can readily be mapped on to one another.

The following year, 1849, Thomson, having finally located a copy of Carnot's original *Reflexions,* published a second paper entitled "An account of Carnot's theory of the motive power of heat, with numerical results deduced from Regnault's experiments on steam." Here Thomson derived a formula for Carnot's function, $C(\theta)$, which determined the work done by a quantity of heat falling through a given temperature. Using Regnault's extensive data he confirmed that $1/C(\theta)$ decreased steadily as the temperature increased; in other words, as Carnot had predicted, other things being equal, a heat engine would be more efficient operating at a lower temperature. It is worth noting in passing how the titles of Thomson's papers indicate the way in which "the precise and extensive experimental data then being published by Regnault were providing a firm basis for the establishment of thermodynamics.... Without Regnault's experimental data thermodynamics would have been impossible" (Cardwell, 1971, p. 240). And, indeed, it is towards the end of the second paper that Thomson coins the label "thermodynamic," although "thermodynamics" as a noun had to wait until 1854.

Meanwhile, William Thomson's confidence in Carnot's theory was being further reinforced. In 1848 his older brother James Thomson (1822–92), a mechanical and marine engineer, had identified an apparent objection to Carnot's conclusions. When water freezes, the "irresistible force" of its expansion can do work, although heat appears to be flowing between water and ice at the *same* temperatures. The only apparent explanation could be that at *increased* pressure the melting point of ice should *decrease;* at first glance this seemed to be highly unlikely. However, in a series of delicate experiments actually performed during his undergraduate natural philosophy lectures at Glasgow in 1850, William Thomson found that such a decrease *did* occur, and by precisely the amount that Carnot's theory required. (And this drop in melting point under pressure is why ice skates are slippery and it is possible to make snowballs.)

By 1850, therefore, William Thomson was strongly committed to Carnot's theory, which was apparently based on the caloric theory of heat or, at least, on an assumption of the conservation of heat. At the same time, however, he was increasingly persuaded of the validity of experiments that demonstrated and measured the quite incompatible conversion of work into heat and vice versa (see chapter 4). From the resolution of this dilemma modern thermodynamics was to emerge (see chapter 5).

Fig. 81.—Lord Kelvin's apparatus for determining the depression of the freezing point of water by pressure. (P.)

Figure 3.6: William Thomson's 1849 apparatus with which he established that, contrary to expectation, the freezing point of ice decreases with increased pressure, thereby confirming Carnot's theories. Edser, Edwin, *Heat for Advanced Students* (London, 1923), fig. 81.

4

THE MECHANICAL EQUIVALENT OF HEAT

INTRODUCTION: WHAT JAMES JOULE DID ON HIS HONEYMOON

I must apologize to the reader, that I have not relieved the tediousness of this paper, by a single brilliant illustration. I have neither propelled vehicles, carriages, nor printing presses. My object has been, first to discover correct principles, and then to suggest their practical development. (Joule, 1839–40, p. 481; Smith, 1998, p. 58)

James Joule's Honeymoon

In the summer of 1847 William Thomson, the precocious new professor of natural philosophy at Glasgow University, went on a touring holiday in the Alps. One day, quite by coincidence, "I met, walking, Mr. Joule, with whom I had recently become acquainted at Oxford. When I saw him before he had no ideas of being in Switzerland (he had even wished me to make some experiments on the temperature of waterfalls) but since that time he had been married and was now on his wedding tour. His wife was in a car [cart], coming up a hill" (Cardwell, 1989, p. 89). James Prescott Joule (1818–89) was a dedicated amateur scientist—and subsequently a lifelong friend—whom Thomson had indeed met for the first time a couple of months earlier at the annual meeting of the British Association for the Advancement of Science (BAAS). In opposition to the traditional material, caloric theory of heat, Joule was passionately committed to the "dynamical" theory that heat was essentially a kind of motion, and that heat and mechanical action could be converted into each other. On this basis he reasoned that the motion of falling water should be converted into heat when the water landed at the bottom of the waterfall; the water at the bottom of the fall should therefore be slightly warmer than the water at the top.

Many years later Thomson recollected that, when they had accidentally met in the Alps, Joule had actually been carrying a thermometer with which he was planning to measure this supposed difference in temperature. Sadly, the thoughts of the new Mrs. Joule are not recorded.

It had been at the same BAAS meeting earlier in the summer that Joule had presented the first substantial account of his "paddle-wheel" experiments on the "mechanical equivalent of heat," experiments designed to measure the amount of heat that could be generated by a given amount of mechanical work. A definitive characteristic of Joule's painstaking experiments had been the use of very sensitive and accurate thermometers to measure very small temperature changes. It is not recorded whether his Alpine measurements were successful. Nevertheless, in retrospect at least, 1847 was very much a turning point in scientific thinking about the relationship between heat and mechanical activity in particular, and in thinking about the relationship between all the various "forces of nature" in general.

Trends in Heat, 1825–50

For most scientists during the second quarter of the nineteenth century, however, the study of "heat" as a scientific topic did not seem especially rewarding. The exciting discoveries in the physical sciences were, they believed, rather to be made in the realms of electricity and magnetism, in electrochemistry, and in organic chemistry. The earlier debates on the nature of heat—whether it might be a material fluid or some form of microscopic motion—seemed sterile and stale. It was slightly accidental, even surprising, that the able French chemist Regnault should have dedicated his life to the study of the thermal properties of fluids.

Nevertheless, there were in this period several strands of enquiry that—with the benefit of hindsight, although maybe *only* with the benefit of hindsight—foreshadowed or eventually contributed to the development of the doctrine of "the conservation of energy" in the 1850s (see chapter 5). There was, for example, an increasing insistence on the "correlation of forces," on the interconnectedness of electricity, magnetism, light, heat, chemical activity, motion, and any other powers or "forces of nature." Despite a lack of enthusiasm for speculative debate about the nature of heat, there was a gradual growth of interest in "dynamical" theories that regarded heat as a form of motion. Most significant, however, was the increasing insistence that there must be a precise, fixed correlation between heat and "work" or "mechanical effect"; although he was not unique, it was Joule who made the definitive contribution in this area with his painstaking measurements of "the mechanical equivalent of heat." Last but by no means least, there was an important contribution from German natural philosophers, prompted partly by the fundamental physiological question of the nature of "life," but having wider implications for the purpose and conduct of physics as a discipline (see Sidebar: "Sense and Sensibility").

But these various strands were not part of a systematic program, targeted on a well-defined problem; they were a very diverse and unconnected clutch of metaphysical, theoretical, and practical investigations. It might be said that, in the renewed interest in dynamical theories of heat and in the determination to understand living organisms in strictly physical terms, there was evidence of a new enthusiasm for a material, mechanical view of natural phenomena. Equally significant was the increasing preoccupation with painstaking and above all precise experimental measurement, a preoccupation conspicuous in the work of Joule as in the work of Regnault, and which came to characterize the experimental aspect of the redefined discipline of physics in the second half of the nineteenth century.

> ### "Sense and Sensibility": The Romantic Reaction Against the Enlightenment
>
> Our meddling intellect
> Misshapes the beauteous form of things –
> We murder to dissect.
> Enough of science and art;
> Close up these barren leaves.
> Come forth, and bring with you a heart
> That watches and receives.
> William Wordsworth (1770–1850),
> *The Tables Turned,* 1798; by "art"
> probably meaning "craft"
> or "technology."

In intellectual history the eighteenth century is often labeled "the Enlightenment." In the aftermath of the Scientific Revolution it was a century that celebrated the power of human reason to understand and manipulate nature and, indeed, society. Such rationalism, increasingly materialist and secular, strongly influenced the reforms of the French Revolution; the eventual excesses of the Terror, among other things, helped to encourage a complex late-eighteenth-century reaction against cold, unfeeling rationality, reaffirming the importance of the imagination and of emotional involvement with the world in order fully to understand it. (Jane Austen's [1775–1817] novel *Sense and Sensibility* [1811] charted the comparable conflict between head and heart, between "sense" and "sensibility," in human affairs.)

This new "Romantic" mentality was influential in literature, music, the visual arts, and architecture, but also in science. (The label "Romantic" ultimately derives from medieval tales of sword and sorcery, called "Romances" because they were written in vernacular Romance languages [e.g., Italian or French] rather than learned Latin; these often fantastical tales enjoyed renewed popularity in the late eighteenth century.) In Germany in particular there emerged a vigorous tradition known as "Naturphilosophie" (nature philosophy), especially as defined by the philosopher F.W.A. Schelling (1775–1854) in his *First sketch of a system of Nature Philosophy* of 1799. In contrast to the Enlightenment tendency to analyze and fragment the world, this tradition insisted on the fundamental interconnectedness and unity of natural phenomena, and a consequent search for underlying pattern. This was rewarding in biology, encouraging comparative anatomy, for example, but could also be stimulating in the physical sciences, encouraging Oersted's search for a connection between electricity and magnetism, for instance.

Although primarily a German movement, the tenets of *Naturphilosophie* did influence some contemporary English scientists, including Davy. The Romantic impulse is most clearly expressed, however, by some English poets of the period; John Keats (1795–1821), like Wordsworth, lamented that "philosophy" (i.e., natural science) would destroy the "awful" (i.e., awe-inspiring) magic of the natural world:

> Do not charms fly
> At the mere touch of philosophy?
> There was an awful rainbow once in heaven:
> We know her woof, her texture; she is given
> In the dull catalogue of common things,
> Philosophy will clip an Angel's wings,
> Conquer all mysteries by rule and line.
> Empty the haunted air, and gnoméd mine—
> Unweave a rainbow.
>
> (Keats, *Lamia*, 1820; in Keats, 1973, p. 431)

This belief that the whole was somehow more than just the sum of its parts was especially significant in the understanding of the nature of life. Was a living organism merely a complex automaton, as Descartes and the mechanical philosophy insisted, or did it contain some extra immaterial spark of vitality? The "vitalist" assertion that organisms, and even organic chemicals, could not be understood entirely in terms of their components was much debated in the early decades of the nineteenth century. It received a vivid and enduring expression in Mary Shelley's (1797–1851) parable of *Frankenstein, or the modern Prometheus* (1818).

From the 1820s onwards, however, there was an increasing backlash against what came to be seen as vague, subjective speculation. The obituary of one eminent German scientist in the prestigious *Annalen der Phyisk* in 1824, for example, recalled his antipathy to "the shallow, superficial treatment of the sciences, the endless hypothesizing, the mystical point of view, and the poetry that had entered science... This mixing of fiction and truth, or poetry and science, the playing with empty, half-true analogies, this guessing and suggesting, instead of knowing and understanding has ruined our good name for us Germans abroad....." For the great German chemist Justus von Liebig (1803–73) *Naturphilosophie* had become the Black Death of science.

THE CORRELATION OF FORCES

> The position which I seek to establish in this Essay is that [any one] of the various imponderable agencies...viz., Heat, Light, Electricity, Magnetism, Chemical Affinity, and Motion,...may, as a force, produce or be convertible into the other[s]; thus heat may mediately [indirectly] or immediately produce electricity, electricity may produce heat; and so on the rest." (Grove, 1846, p. 8; Kuhn, 1977 [1959], p. 79)

New Correlation Phenomena

Towards the end of the eighteenth century, the phenomena investigated by experimental natural philosophy—the various "forces of nature," such as electricity, magnetism, heat, and chemical affinity, for example—were most commonly explained in terms of the action of "subtle fluids" such as "caloric." Here "forces" was used in a broad, qualitative sense, rather than the precise, quantitative sense of Newtonian mechanics. The "fluids" were normally considered to be "imponderable" (i.e., weightless), and to be composed of particles that were mutually repulsive, but attracted to particles of ordinary matter. Initially, distinct fluids were proposed for separate types of phenomena—a fluid for heat, one (or maybe two) fluids each for electricity and magnetism, and so on—although later the various fluids tended increasingly to be regarded as but different manifestations of a single fluid.

From 1800 onwards there did indeed emerge a number of striking new experiments that clearly connected previously separate "forces"—electricity with chemistry, electricity with magnetism, heat with light, light with magnetism. Most remarkable perhaps was Alessandro Volta's (1745–1827) 1800 construction of the bimetallic "Voltaic pile" to produce current electricity. Once the identity between the new current electricity and the old electrostatic electricity had been established, this phenomenon clearly linked electricity and chemistry. This link was rapidly and spectacularly confirmed by the work of Davy and others on the electrolytic decomposition of water and other substances, resulting in the isolation of a number of new chemical elements, such as sodium and potassium. The connection was given greater, quantitative precision by Michael Faraday's (1791–1867) subsequent discovery in the 1830s of his laws of electrochemistry. Meanwhile, the long-suspected link between electricity and magnetism was confirmed in 1820 by Hans Christian Oersted (1777–1851), when he noted the deflection of a compass needle by an electric current. This link was reinforced through the 1820s and 30s by Faraday's construction of electromagnetic motors and his demonstration of the induction of electric currents by changing magnetic influence. The connection between heat in its radiant form and light has been discussed above. That polarized light could be influenced by magnetism was demonstrated by Faraday in 1845. So it is not surprising that in 1834 Mary Somerville (1780–1872) entitled her popular scientific textbook *On the Connexion of the Physical Sciences*.

The Unity and Conversion of Force

Such manifold connections inevitably suggested the concept of some more fundamental underlying and unifying "force," a concept that was already embedded in general terms in the German tradition of *Naturphilosophie*. The German natural philosopher G. F. Mohr (1806–?79), for instance, asserted in 1837 that "besides the known 54 chemical elements, there is, in the nature of things, only one other agent, and that is called force; it can appear under various circumstances as motion, chemical affinity, cohesion, electricity, light, heat, and magnetism, and from any one of these types of phenomenon all the others can be called forth" (quoted in Kuhn, 1977 [1959], p. 78). Belief in the mutual convertibility of different forces of nature was a fundamental guiding influence on the direction of Faraday's researches. In his paper on magnetism and light he affirmed that "the various forms under which the forces of matter are made manifest have one common origin; or, in other words, are so directly related and mutually dependent, that they are convertible, as it were, one into another, and possess equivalents of power in their action" (Harman, 1982, p. 35). The *Correlation of physical forces* (1846) by William Robert Groves (1811–96), as quoted at the head of this section, presented a popular account of such views.

These assertions were often claimed retrospectively as expressions of "the conservation of energy" that was explicitly formulated in the 1850s (see chapter 5). Clearly there were similarities, but there were also important differences. Arguably, a profound metaphysical and methodological shift was involved, and Faraday himself was to express reservations about the new science of energy in the late 50s. But more specifically, in the 1840s it was not at all clear whether the conversion of one force into another was a precise, quantitative exchange, nor how to measure it. Thus Grove noted that if motion were "subdivided or changed in character, so as to become heat, electricity, etc.; it ought to follow, that when we collect the dissipated and changed forces, and reconvert them, the initial motion, affecting the same amount of matter with the same velocity, should be reproduced, and so of the change of matter produced by the other forces" (Grove, 1846, p. 47; Kuhn, 1977 [1959], p. 80). And he went on to suggest that "the great problem that remains to be solved, in regard to the correlation of physical forces, is the establishment of their equivalent of power, or their measurable relation to a given standard."

With hindsight, it might seem obvious that mechanical effect or "work" would provide an appropriate measure for such "equivalents of power." A glance at some of the suggestions canvassed by exponents of the "conversion of forces" should dispel this assumption, however. Grove saw the key link in Dulong and Petit's law connecting specific heats and chemical combining weights, whereas Mohr sought a connection between the heat needed to raise the temperature of water and the force needed to re-compress it to its original volume. Clearly it was not at all obvious to the scientists of the day that the "standard" required could be found in consideration of the output of heat engines. This crucial connection emerged from the largely separate domain of power engineering, and most forcefully from the work of James Joule. Before examining Joule's work on this problem, however, it will be useful to survey briefly the resurgence of dynamical theories of heat, especially in England, in the second quarter of the nineteenth century.

THE EARLY RESURGENCE OF DYNAMICAL THEORIES OF HEAT

The mechanical equivalents of heat determined by the various series of experiments given in this paper...afford a new and, to my mind, powerful argument in favour of the dynamical theory of heat which originated with Bacon, Newton, and Boyle, and has been at a later period so well supported by the experiments of Rumford, Davy, and Forbes. With regard to the detail of the theory, much uncertainty at present exists. (Joule, 1845, p. 381; Smith, 1998, p. 70)

Dynamical Theories in General

The question of the true nature of heat had been largely set to one side in the second quarter of the nineteenth century. Most textbooks of the period gave fairly even-handed and inconclusive accounts of the relative strengths and weaknesses of the material (caloric) and dynamical (motion) theories of heat. Even so, there does seem to have been some increase in interest in dynamical theories in the 1830s and 40s. We have already noted Sadi Carnot's conversion to a dynamical theory in his late, unpublished manuscript writings of the early 1830s. There were at least another half dozen proponents of the theory in the remainder of the two decades, most of whom were, like Carnot, relatively marginal figures, although this was far from true of the eminent French physicist André Marie Ampère (1775–1836), who published an influential version of the dynamical theory in 1832.

There was a wide variety of dynamical theories: Given that heat was essentially a form of motion, it still remained to establish exactly what was moving and what kind of motion was involved—whether vibration, rotation, or translation. Ampère, for instance, believed that the ultimate molecules of all substances (including the elements) were polyatomic; heat, therefore, consisted in the relative *vibrations* of the atoms *within* these molecules, and heat was transmitted not directly but by the communication of this vibration through the surrounding aether. This theory, often labeled the "wave theory of heat," was quite well received. The recently established identity of radiant heat and light was clearly influential. In 1837 Mohr developed a theory similar to Ampère's, but he added some interesting suggestions about the significance of an "absolute zero" of temperature and the invalidity of the concept of "latent heat"; of particular interest in the light of later developments, he also asserted that the difference between the two principal specific heats of gases was due to the work done in expansion (see below). In an 1845 essay on "Physiological Heat," the German physiologist and physicist Hermann von Helmholtz (1821–94), whose later work we shall discuss in more detail below, rejected the caloric theory in favor of a dynamical explanation, but without committing himself to any particular detailed model.

Whereas the heat of solids and liquids was generally attributed to some form of vibration, whether between or within molecules, explanations for the behavior of gases varied more widely. Broadly speaking, there were two main approaches. The first approach in many ways reworked the traditional, Newtonian model of an essentially *static* lattice of gas particles, held apart by mutual repulsion. Traditionally this repulsion had been attributed to atomic atmospheres of self-repelling caloric; in the new dynamic theories the repulsion was now generated by the rotation or vibration of an atmosphere, often composed of an "electric fluid." The second approach treated gas pressure (and temperature) as a consequence of the bombardment of the walls of any container by rapidly moving particles of gas, much in the manner of the modern "kinetic" theory of gases. Interest in this approach was extremely limited before the mid-1850s, being largely restricted to a couple of marginal British natural philosophers.

A theory of the first type was advanced by Joule in 1844. Joule maintained that his experimental results provided "a new and, to my mind, powerful argument in favour of the dynamical theory of heat." Ingeniously developing a "beautiful idea" of Davy's about the rotational motion of gas molecules in the light of Faraday's more recent researches, Joule supposed that all atoms were surrounded by "atmospheres of electricity [which] revolve with enormous rapidity round their respective atoms; that the momentum of the atmospheres constitutes 'caloric,' while the velocity of their exterior circumference determines what we call temperature...the centrifugal force of the revolving atmospheres is the sole cause of expansion on the removal of pressure" (Joule, 1846, pp. 110–111; Smith, 1998, p. 70). Radiant heat could be explained on the assumption that the atmospheres possessed "the power of exciting isochronal undulations in the aether which is supposed to pervade space."

The Kinetic Theory of Gases

By the end of the 1840s, however, Joule himself expressed a preference for the second "kinetic" approach to the structure of gases. The basic theory had a reputable pedigree going back at least to Daniel Bernoulli in 1738. However, just as dynamical theories in general had been a distinctively British preoccupation in the early nineteenth century, so was interest in the kinetic explanation of the properties of gases in the second quarter. (Perhaps the image of the random jostling of molecules appealed to the densely urban British culture of laissez-faire individualism.) Even in Britain, however, the two chief protagonists of the theory remained fairly marginal members of the scientific community.

John Herapath (1790–1868) has been described as a typical English eccentric. He was a man of relatively humble origins whose natural abilities enabled him nevertheless to obtain a reasonable education. His 1821 paper "A Mathematical Inquiry into the Causes,... of Heat,..." developed a kinetic theory of gases in the style of Bernoulli, but taking into account the recent contributions of Gay-Lussac and Dalton. Herapath derived the equation

$$PV = 1/3Nmv^2$$

connecting the pressure (P) and volume (V) of a gas with the number (N), mass (m), and velocity (v) of its particles, but he also developed further interesting arguments about diffusion, change of state, adiabatic heating, and the existence of an absolute zero of temperature: "the degree therefore of absolute cold is where the particles have no motion" (Herapath, 1821, p. 303; Cardwell, 1971, p. 148).

Despite earlier correspondence with Davy, Herapath's paper had been rejected by the Royal Society before its eventual publication in the less prestigious *Annals of Philosophy*. Some of his conclusions were perhaps idiosyncratic—the claim, for example, that "true" temperature was proportional to molecular velocity or momentum rather than to the velocity squared or *vis viva*—but the main reason for official coolness was probably the very antiquated presentation in a Newtonian style of Euclidean geometry and proportions. Herapath nonetheless continued to promote his theories, later through the pages of the *Railway Journal*, of which he had fortunately become the editor. An extended presentation of his ideas in 1847 in a two-volume book *Mathematical physics; or the mathematical principles of natural philosophy: with a development of the causes of heat, gaseous elasticity, gravitation, and other great phenomena of nature*—the title is presumably an ambitious echo of Newton's *Principia Mathematica*—came to the attention of Joule. Consequently, by 1848 Joule had come to consider that Herapath's kinetic theory was "somewhat simpler" than his own "hypothesis of a rotary motion" (Joule, 1884, i, p. 294; Brush, 1976, p. 161).

An even more engaging figure is John James Waterston (1811–83), a well-educated Scottish engineer who spent a significant part of his working life as an instructor in navigation and gunnery to cadets of the East India Company in Bombay. In 1845 he submitted to the Royal Society a perfectly clear account of the kinetic theory of gases, which included a number of interesting contributions, such as a statement of the equipartition theorem and a calculation of the ratio of the two principle specific heats of gases. The main academic referee, however, asserted that "the paper is nothing but nonsense, unfit even for reading before the Society"—although by the time this report was received the paper had already been read to the Society, and an abstract was duly published (see Waterston, 1893, p. 2). Waterston made several further attempts to publicize the kinetic theory in the 1850s, especially through meetings of the BAAS, but was almost entirely ignored. It is perhaps not surprising that in later life, according to his nephew, "He talked in a manner that seemed to me strangely contemptuous of scientific men with but few exceptions...any mention of [learned societies] generally brought out considerable abuse..." (Rayleigh, 1924; Brush, 1976, p. 143).

This apparent official antipathy suggests that the kinetic theory in particular had acquired a reputation as an excessively speculative—even crank—hypothesis. Despite the evident growth of interest by the late 1840s, even dynamical theories of heat in general remained largely speculative and marginal. Only in the 1850s, in the aftermath of the acceptance of Joule's equivalence of heat and mechanical effect (see chapter 5), did scientific opinion quite rapidly crystallize in their favor—albeit still in very varied and often quite general terms (see chapter 6).

JOULE AND THE MECHANICAL EQUIVALENT OF HEAT

Since constant practice had enabled me to read off with the naked eye to 1/20th of a division, it followed that 1/200th of a degree FAHR. was an appreciable [measurable] temperature. (Joule, 1850, p. 64)

Early Researches: From Ingenious Invention
to Philosophical Experiment

James Prescott Joule (1818–89) was the son of a wealthy Manchester brewer who had prospered as the town, the archetypal Industrial Revolution cotton-mill boomtown, expanded prodigiously in the late eighteenth and early nineteenth centuries. Educated largely by private tutors, including Dalton, Joule was able to devote himself to a gentlemanly life of amateur natural philosophical investigation. Stimulated by an early interest in the commercial potential of the new electric motor, he moved rapidly to a scientific study of the distribution of heat, chemical action, and mechanical activity around an electric circuit. The realization by 1843 that the heat and work were invariably "proportional" led to increasingly sophisticated and precise experiments to determine the "mechanical equivalent of heat," firstly in 1844 in the case of the compression and expansion of air, and subsequently in the case of fluid friction. The response of the scientific community remained at best lukewarm until Thomson's attendance at the BAAS in Oxford in 1847.

The possibility of constructing electric motors (powered by voltaic electric batteries) and "magneto-electric machines" (that is, generators or dynamos to generate electric current) stimulated a vigorous practical and popular culture of design and display of electrical devices. In 1838 the young Joule promptly entered this world of ingenious invention with his "Description of an Electro-magnetic Engine" published in the *Annals of Electricity, Magnetism and Chemistry; and Guardian of Experimental Science.* His machine was described by the editor as an "exceedingly ingenious arrangement of electromagnets of soft iron," in which the "bars…are of a peculiar construction, and the transposition of their polarity effected by an exceedingly ingenious contrivance. Mr. Joule proposes to apply his engine both to locomotive carriages and to boats" (Smith, 1998, p. 58). However, the performance of Joule's electrical engine, albeit ingenious, was disappointing. The electrical engine was envisaged, after all, as a potential rival to the steam engine, so economy was crucial. Tests conducted by Joule a couple of years later seemed to confirm that the "mechanical force" to be expected from a battery was of the order of 331,400 foot-pounds for every pound of zinc consumed, which was deemed to compare poorly with the 1,500,000 or so foot-pounds that a good steam engine could now commonly produce from a pound of coal.

Consequently, he resolved to devote himself to a more systematic study of the principles underlying the effective operation of the electric motor and generator. This was a step towards the elite domain of the experimental philosopher, away from the popular world of commercial invention and spectacle. Joule's objective was to understand the principles and constraints influencing the output of various effects in an electric circuit. Initially his attention focused on electrical resistance, resulting in an 1840 paper "On the Production of Heat by Voltaic Electricity," which outlined the now well-known law that the heating effect of a circuit is proportional to the resistance and to the square of the current, i.e., power, $W = i^2R$.

As far as a straightforward combination of batteries and resistances was concerned, Joule was able to assert in an early 1843 paper "On the Heat

Evolved During the Electrolysis of Water" that "the whole caloric of the circuit is exactly accounted for by the whole of the chemical changes" (Joule, 1846 [1843], p. 103; Smith, 1998, p. 64). But Joule was also interested in more complex arrangements involving not just batteries and resistances but also electric motors and generators. Thus in the same paper Joule asserted that a "magnetic electrical machine" (or generator) "enables us to convert mechanical power into heat, by means of the electric currents which are induced by it" (Joule, 1846 [1843], p. 104–5). It was possible to object, however, that the generator was not the source of the heat but was merely *transferring* caloric from one part of the circuit to another. Maybe as the resistance warmed up, the generator itself cooled down, just as the chemical power of a battery was supposed to be reduced in the production of heat?

To resolve this problem Joule conducted experiments on a generator immersed in a water-bath. If heat were merely transferred from one part of the circuit to another, there should have been no overall increase in the temperature of the bath. An increase in temperature *was* detected, however, confirming that "*heat* is *generated* by the magneto-electric machine,..." (Joule, 1846 [1843], p. 435; Smith, 1998, p. 65) and that "the mechanical power exerted in turning a magneto-electric machine is *converted into the heat* evolved by the passage of the currents of induction through its coils;..." (Joule, 1884, i,], p. 172; Cardwell, 1971, pp. 232–33).

In itself this was a satisfying result. But Joule's generator had been powered by descending weights attached by lengthy threads to its axle. Thus he was able not only to measure the amount of heat generated but also to calculate, from their sizes and the distances descended by the weights, the mechanical input or work done to drive the generator. Comparison of the two figures gave an exact measure of the "proportion" between the heat and the work. At the 1843 meeting of the BAAS in Cork, Ireland, Joule read a paper describing these experiments entitled "On the Calorific Effects of Magneto-electricity, and on the Mechanical Value of Heat." He deduced that "the quantity of heat capable of increasing the temperature of a pound of water by one degree of Fahrenheit's scale is equal to, and may be converted into, a mechanical force capable of raising 838 lb. to the perpendicular height of one foot" (Joule, 1843, p. 441; Smith, 1998, p. 66). (This result corresponds to an equivalence of 4.51 joule per calorie.) He concluded by promising to "lose no time in repeating and extending these experiments, being satisfied that the grand agents of nature are, by the Creator's fiat, *indestructible;* and that wherever mechanical force is expended, an exact equivalent of heat is *always* obtained" (Joule, 1843, p. 442; Smith,1998, p. 68).

The Mechanical Equivalent of Heat,1: The Condensation and Rarefaction of Air

From the summer of 1843 onwards the emphasis of Joule's research shifted significantly. Up to this point his experiments had remained oriented towards his original interest in the efficiency of battery-operated electrical machines, the conversion of electrochemical action into work and heat. The preceding quotation suggests that he now realized that his studies had a more fundamental, even theological, significance, and from this time onwards also he more commonly connected his researches with establishing the dynamical theory of heat. Following the lead of his latest paper, he now focused more narrowly and directly on the interconversion of heat and work, and—with one major exception, which will be discussed immediately—especially on the conversion of mechanical action into heat through friction.

Before turning his attention to the development of frictional methods of measuring the mechanical equivalent of heat, Joule tried a different approach. He argued that the adiabatic cooling (or heating) of a gas must be due, not to a supposed "latent heat of expansion," but rather to the work done by (or on) the gas in expanding (or being compressed); a comparison of the heat absorbed (or generated) with the work done would provide another route to the measurement of their equivalence. This approach was not entirely new, but Joule seems to have been the first to make a direct experimental measurement, and in an attempt to investigate directly the issue of an alleged "latent heat of expansion."

The possible significance of adiabatic phenomena had been noted by a number of scientific and engineering writers in the preceding decade. Mohr, for example, in his 1837 papers on the nature of heat, included the observation that the difference between the specific heat of a gas at constant volume and the specific heat at constant pressure (when the gas would expand as it was heated) should be explained by the work done by the gas, rather than in terms of a "latent heat of expansion." A similar point was made by Julius Robert Mayer (1814–78) in 1842, with the bonus of an explicit calculation of the implied rate of conversion or equivalence between the heat and the work: Using available data on the ratio of the two specific heats, he estimated that "the warming of an equal weight of water from 0°C to 1°C corresponds to the fall of an equal weight from the height of about 365 metres" (Mayer, 1862 [1842], pp. 371–7; Smith, 1998, p. 75), corresponding to a value of 3.58 joules per calorie for the mechanical equivalent of heat. In 1845 German schoolteacher Karl Holtzmann (1811–?) performed much the same calculation in a paper "On the Heat and Elasticity of Gases and Vapours and on the Principles of the Theory of Steam Engines," arriving at a figure 374 m (equivalent to 3.67 joules per calorie). Intriguingly, it was Clapeyron's paper that had stimulated Holtzmann's speculations and, although ultimately agnostic about the true nature of heat, he stuck to the *conservation* of heat as a principle, if not explicitly to a material caloric. For Holtzmann, therefore, the connection between heat and work done was based on Carnot's ideas: The figure 374 m provided a *measure* of heat in terms of work, *not* a conversion factor.

In 1844 Joule was almost certainly unaware of these initiatives. His much more incisive approach was to make a direct empirical measurement of the work done and the heat generated, and thus of the conversion coefficient. He presented his findings once again to the Royal Society, in a paper entitled "On the Changes of Temperature Produced by the Condensation [compression] and

Rarefaction [expansion] of Air"; the Society once again contented itself with publishing an abstract, which recorded that Joule had

> contrived an apparatus where both the condensing-pump and the receiver were immersed in a large quantity of water, the changes in the temperature of which were ascertained by a thermometer of extreme sensibility [sensitivity]. By comparing the amount of force expended in condensing air in the receiver with the quantity of heat evolved...it was found that a mechanical force capable of raising 823 pounds to the height of one foot must be applied in the condensation of air, in order to raise the temperature of 1 lb. of water 1° of Fahrenheit's scale. (Joule 1884, i, p. 171; Smith, 1998, p. 68)

This result was satisfyingly close to that obtained by a very different method the year before.

Joule also tackled a major objection to this procedure, initiating a line of investigation that was to be important for the later development of the kinetic theory of gases. It was not at all certain that there was no intrinsic latent heat of expansion, which would result in a fall in temperature *whether or not* external work was done. Once again Joule conducted a direct investigation, somewhat reminiscent of Gay-Lussac's much earlier (and cruder) experiment using twin flasks in an attempt to measure the relative specific heats of gases: "In another experiment, when air condensed [compressed] in one vessel was allowed to pass into another vessel from which the air had been exhausted [evacuated], both vessels being immersed in a large receiver full of water, no change of temperature took place, no mechanical power having been developed" (Joule, 1884, i, p. 171; Smith, 1998, p. 69). According to Joule, therefore, it followed that the fall in temperature detected in the first experiment must be due to the external "mechanical power" that was developed.

The amounts of heat and work involved in these air experiments, and the corresponding temperature changes, were quite small and difficult to measure. It is perhaps for this reason, or maybe because of the tepid official reception, that Joule subsequently turned his attention to liquids and solids, pursuing an approach that was closer to his previous experiment on the generation of heat by an electric dynamo.

The Mechanical Equivalent of Heat, 2: The Paddle-wheel Experiments

Joule's most famous and influential sequence of experiments, progressively refined between 1845 and 1850, involved the generation of heat by fluid friction. Liquid in a calorimeter was set in motion by an externally driven paddle-wheel; as the liquid came to rest, according to Joule, the force of motion must be converted into heat and cause a small rise in temperature. He presented accounts of these experiments at the BAAS meetings in Cambridge in 1845 and in Oxford in 1847, but the definitive account "On the Mechanical Equivalent of Heat" was printed in the Royal Society's *Philosophical Transactions* in 1850.

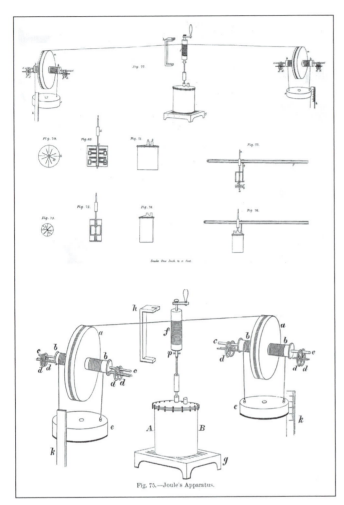

Fig. 75.—Joule's Apparatus.

Figure 4.1: James Joule's apparatus, first developed in 1845, to measure the mechanical equivalent of heat; the falling weights (e) drive paddlewheels that churn the water inside the copper calorimeter (AB), causing a very slight but measurable rise in temperature. Preston, *Theory of Heat* (London, 1904), fig.75.

In the main sequence of experiments, the copper vessel was filled with water that warmed up very slightly as the paddle-wheel rotated. The mechanism that drove the rotating paddle-wheel was similar to that employed in the 1843 submerged-dynamo experiment, a pair of heavy lead weights, each up to 29 pounds, descending through some five feet, thus enabling the effort used to turn the paddle-wheel to be calculated and compared with the heat generated.

The rise in temperature at each descent of the weights was only about 1/40th degree Fahrenheit (i.e., about 1/80th degree Celsius). Joule therefore repeated the descent of the weights 20 times for each complete "run" of the experiment; even so, the temperature rise for each run was only about 1/2°F. Joule took very considerable precautions to reduce heat exchange with the surroundings, including "a large wooden screen...[that] completely obviated the effects of radiant heat from the person of the experimenter," and made dummy runs of the experiment without the weights in order to measure the spontaneous change in temperature of the water over a similar time period. He also repeated the whole procedure up to 40 times. But whatever precautions were taken, it was clear that the accuracy and precision of the thermometers used was all-important. Joule was able to combine the latest standardizing procedures developed by Regnault with the technical abilities of local Manchester instrument maker Benjamin Dancer (1812–87):

> The thermometers employed had their tubes calibrated and graduated according to the method first indicated by M. Regnault. Two of them, which I shall designate by A and B, were constructed by Mr. Dancer of Manchester;...The graduation of these instruments was so correct, that when compared together their indications coincided to about 1/100th of a degree Fahr[enheit]....And since constant practice had enabled me to read off with the naked eye to 1/20th of a division, it followed that 1/200th of a degree Fahr. was an appreciable [measurable] temperature. (Joule, 1850, p. 64)

It has been suggested that the importance of delicate temperature control within the family business of brewing gave Joule some of the attitudes and skills that he required.

For the 1850 paper, Joule conducted a similar experiment using mercury instead of water and another experiment to measure the heat generated by the solid friction of cast iron. It remained the result for water with which he was most satisfied: "I consider that 772.692, the equivalent derived from the friction of water, is the most correct, both on account of the number of experiments tried, and the great capacity of the apparatus for heat" (Joule, 1850, p. 82). This figure (an equivalence in modern terms of 4.157 joule/calorie) compared well with the results previously reported to the BAAS, namely: "781.5, 782.1 and 787.6 respectively, from the agitation of water, sperm-oil and mercury" (Joule, 1850, p. 64). (Sperm oil, obtained from sperm whales, was widely used for domestic lighting in Victorian Britain, greatly to the detriment of whales.) Joule therefore concluded:

> 1st. *That the quantity of heat produced by the friction of bodies, whether solid or liquid, is always proportional to the quantity of force expended.* And,
>
> 2nd. *That the quantity of heat capable of increasing the temperature of a pound of water (weighed* in vacuo, *and taken at between 55° and 60°) by 1° FAHR., requires for its evolution the expenditure of a mechanical force represented by the fall of 772 lbs. through the space of one foot.* (Joule, 1850, p. 82)

For Joule, the establishment of a precise, universal "mechanical equivalent of heat" proved the dynamical theory of heat, that is to say, proved that heat was indeed a form of motion.

The Reception of Joule's Work

The publication of this paper in 1850 by the Royal Society meant that Joule had finally achieved the recognition by elite scientific professionals for which he had long been struggling, and he was very soon elected a fellow of the Society. The tide had turned for Joule in 1847 when he met William Thomson at the Oxford meeting of the BAAS. Thomson was impressed by Joule's experiments, but by no means entirely convinced of his broader interpretation of his results, commenting that Joule was "wrong in many of his ideas, but he seems to have discovered some facts of extreme importance, as for instance that heat is developed by the frict[io]n of fluids in motion" (letter to his father, quoted in Smith, 1998, p. 79). For many scientists, the very reality of the frictional heating of water (and other fluids) remained suspect—the idea that one could heat water simply by vigorous stirring or shaking seemed absurd. Thus, on his return to Glasgow, rather than attempting to replicate Joule's precision measurements, Thomson set about creating vivid practical demonstrations of the basic effect, rather in the tradition of Rumford. Using a manually driven paddle-wheel—subsequently converted to be powered by the machinery at a cotton-mill—he achieved undeniable temperature rises, although he failed in his objective of getting the water to boil. Joule was delighted, and he wrote to Thomson: "I had found considerable difficulty in persuading our scientific folks here that the heat derived in my experiments was not derived from the friction of the bearings under water, but your experiments were not to be cavilled at. I have also always found a difficulty in making people believe that fractions of a degree could be measured with any great certainty, but your experiments showing a rise of temperature of 30° or 40° would prove the truth of the fact by the warmth as felt by the hand" (Smith, 1998, p. 81).

Yet Thomson remained cautious about fully accepting Joule's increasingly confident belief that heat and "mechanical effect" were always straightforwardly convertible one into the other. The theory—and most of the calorimetric practice—of heat in the first half of the nineteenth century depended on the *conservation* of heat, even if not upon a material "caloric." A letter from Thomson to Joule in 1848 made it clear that even now for Thomson the "quantity of heat" in a body in a given state must be fixed, a "state function," like pressure or temperature. It took another three or four years of vigorous private debate between Joule, Thomson, and his associates to integrate Joule's findings into the mainstream academic study of heat (see chapter 5). Meanwhile, Joule's experiments had also been noticed in Germany.

LIVING FORCE *(VIS VIVA)* AND LIFE FORCE *(LEBENSKRAFT)*: PHYSICS AND PHYSIOLOGY IN GERMANY

Whence, I often asked myself, did the principle of life proceed? . . . I determined thenceforth to apply myself more particularly to those branches of natural philosophy which relate to physiology.

—Mary Shelley, *Frankenstein*

During the first half of the nineteenth century the German university system underwent a fundamental reform and expansion. Physiology was one of the subjects that particularly flourished in this period, the German universities acquiring an international reputation for their experimental investigations. It is not accidental that Mary Shelley chose a German setting for *Frankenstein*, her story of extreme spare-part surgery. Much debated—the Frankenstein story reflects this—was the question of whether living organisms, and "life" itself, could be explained in terms of complex but purely material mechanisms, or whether they could only be understood in terms of some distinctive, additional "life force" (*Lebenskraft*). Against this at first sight unlikely background emerged the two major German contributions to the study of heat before 1850. Although these studies tackled similar problems, their authors adopted markedly different approaches and certainly enjoyed very different degrees of success in their careers.

Mayer "on the Forces of Inorganic Nature"

In contrast to James Joule, Julius Robert Mayer never really received the official recognition that arguably he deserved. Although his work lacked Joule's experimental precision and thoroughness, in 1842 he was nevertheless the first person to assert that there must be an exact numerical equivalence between heat and work, and to offer a valid estimate of its value. He also combined this insight with a vision broader than Joule's of the interconnectedness of different "forces."

Mayer trained as a medical doctor and continued to practice as one for most of his life. Much in the eighteenth-century tradition—and in stark contrast to Joule and Thomson's industrial interests—Mayer arrived at his insights from a *physiological* starting point. Shortly after qualifying as a physician, he took a job as a ship's doctor that took him to the tropics. There he noticed that venous blood—normally in his experience in a temperate climate much darker—remained almost as bright red in color as arterial blood. The explanation evidently lay in the greater warmth of the tropics and the consequently lower level of metabolic activity needed to keep the body warm. However, it struck him that, in order to account fully for a body's metabolic activity, one should consider not only the internal *heat* produced but also the amount of external *work* that was being done. More work and less heating would produce the same overall metabolic effect as less work and more heating: Work and heat were thus interchangeable and, in a sense, equivalent.

This perception of the physiological equivalence of heat and work led Mayer to some rather metaphysical speculations—possibly influenced by the *Naturphilosophie* tradition—and to one specific and original proposition. His general speculations led him to an affirmation of the overall convertibility and "indestructibility of force," very much in the Faraday tradition, although Mayer did not in fact accept the dynamical theory that heat *was* motion: It was for him a distinct, independent "force." His original contribution was to propose that the equivalence between heat and work could be given a precise numerical value. Moreover, this value could be calculated, he suggested, by considering the descent of a column of mercury in a tube as it compressed air trapped at the end of the tube. A comparison of the work done by the descent of the mercury with the heating up of the air would give a value for the heat-work equivalent. Using available data on the ratio between the two principal specific heats of air, Mayer calculated that "the warming of an equal [given?] weight of water from 0° to 1°C corresponds to the fall of an equal weight from the height of about 365 metres" (Mayer, 1862 [1842], p. 377; Brush, 1965, i, p. 77). Unfortunately, he did not initially publish the details of his calculations.

Moreover, his argument assumed that the internal energy of a gas was constant, that a free expansion without doing any work would result in no change in temperature. It was precisely this assumption, subsequently labeled "Mayer's hypothesis" by Thomson, that Joule's later twin-cylinder expansion experiment was designed to test, although Mayer was able to point to the similar, earlier—and much cruder—experiment by Gay-Lussac.

Partly because of the flaws in his argument, but much more because of the generally speculative tone and the absence of any original experimental evidence, Mayer's paper was completely ignored. In 1848, when he became aware of Joule's published work on the mechanical equivalent of heat, Mayer wrote to the French Academy of Sciences pointing out his own earlier contributions. Joule's not unreasonable response, as he wrote to Thomson in 1851, was that "I shall be quite content to leave M. Mayer in the enjoyment of the credit of having predicted the law of the equivalency. But it would certainly be absurd to say that he has established it" (Smith, 1998, p. 74.). This seems to be a fair enough assessment. Some later British writers on the history of thermodynamics seem to have gone out of their way to belittle the significance of Mayer's contribution; this trend is very largely due to the excessively jingoistic accounts of Thomson's Scottish colleague Peter Guthrie Tait (1831–1901), which sought to establish the discovery of the conservation of energy as a primarily British achievement, exalting the reliability of Rumford and Davy's experiments, refusing to admit any compromise of Joule's pre-eminent role, and thus seeking to denigrate Mayer—and minimizing the importance of Helmholtz and Clausius to boot.

Helmholtz "on the Conservation of Force"

Hermann Ludwig Ferdinand von Helmholtz (1821–94) shared much of Mayer's background and agenda, but in his methods and, above all, in his academic success he could hardly have been more different. Like Mayer, Helmholtz initially trained as a medical doctor—although his early interest was in physics, there were still few established avenues for training as a physicist—and he served several years as a military surgeon. Thereafter he became a professor of physiology at a succession of German universities, before becoming an illustrious professor of experimental physics at Berlin in 1871. In the 1840s, however, while quite unaware of Mayer's work, Helmholtz was fundamentally concerned to eliminate from physiological thinking the Romantic *Naturphilosophie* notion of a distinct "life force" (*Lebenskraft*) characteristic of living organisms, a "force" that could not be reduced to other ordinary material forces. Thus, in 1845 he suggested that "one of the most important questions of physiology, one immediately concerning the nature of the *Lebenskraft*,...[is] whether the life of organic bodies is the effect of a special self-generating, purposive force or whether...the mechanical force and heat generated in the organism can be completely deduced from the process of material exchange or not" (Smith, 1998, pp. 129–30). He developed his ideas at length in an 1847 paper "On the Conservation of Force," which—like Mayer's early efforts, in fact—was rejected for publication by the prestigious scientific journal Poggendorff's *Annalen* and had to be published independently. Helmholtz' main argument was that, if there were some such special "vital" force, then it would be possible to generate something for nothing; the output of living organisms would be greater than the initial purely material input, and they would become some kind of "perpetual motion machine." If on the other hand the output were the same, what contribution or difference did the special force make?

Beyond this general criticism of "vitalism," Helmholtz went on to sketch a comprehensive, mathematical picture of how the world must be constructed if, as was by then generally accepted, a "perpetual motion" machine were impossible; if, in other words, the overall quantity of "force" in the world were constant or conserved. Broadly speaking, he adopted a Laplacian or "astronomical" view that all natural phenomena could be understood in terms central forces—gravitational, electrical and so on—acting between point particles. "The problem of physical science," he insisted, "is to reduce natural phenomena to unalterable forces of attraction and repulsion, whose intensity depends on the distance" (Helmholtz, 1853 [1847]; Brush, 1965, i, p. 92).

To this system he applied the principle of the conservation of mechanical energy. The eighteenth-century principle of the conservation of *vis viva* had been developed in the early decades of the nineteenth century into a more general principle of the conservation of mechanical energy, maintaining that the sum of the kinetic energy and the potential energy of a body moving under the influence of a central force would be constant. As it swings to and fro under the influence of gravity the bob of a pendulum continually exchanges its kinetic energy of motion for potential energy of extra height, and vice versa. Thus, although *vis viva* alone was *not* necessarily constant, the combination of the *vis viva* and the potential energy of a body *was*. Helmholtz generalized an idea normally restricted to the motion of weights under the influence of gravity. He did not use the terms "kinetic energy" and "potential energy," but referred instead to "vis viva" and to "tensions" or "stretch-forces" (*Spannkräfte*), with the satisfying association of the energy stored in a spring:

> In all cases of the motion of free material points under the influence of their attractive and repulsive forces, whose intensity depends solely upon the distance, the loss of tension [potential energy] is always equal to the gain in *vis viva* [kinetic energy], and the gain in the former equal to the loss in the latter. Hence *the sum of the existing tensions and vires vivae* [kinetic energies] *is always constant*. In this most general form we can distinguish our law as *the principle of the conservation of force*. (Helmholtz, 1853 [1847], pp. 124–5; Smith, 1998, p. 134)

Since the "force" of any possible natural agent was therefore necessarily conserved, the total "force" of the material world must be constant, perpetual motion was indeed impossible, and no special "vital force" was required (or, indeed, admissible) to account for the behavior of living organisms. Physiology had to be ultimately a branch of physics.

Helmholtz recognized, however, that there were many situations—inelastic collisions, friction—where *vis viva* was neither conserved nor smoothly exchanged with (and recoverable from) some "potential" store. Whenever *vis viva* was lost, however, some other "force," such as heat or electrostatic charge, seemed to appear. Helmholtz's mathematics provided no insight into the relationship between *different* forces, however, so there remained the *empirical* question "whether the sum of these [new] forces always corresponds to the mechanical force which has been lost," and whether (temporarily ignoring possible chemical and electrical effects)

> for a certain loss of mechanical force a definite quantity of heat is always developed, and how far can a quantity of heat correspond to a mechanical force... For the solution of the first question but few experiments have yet been made. Joule has measured the heat...developed in vessels in which the water was set in motion by a paddle-wheel;... His method of measurement however meets the difficulty of the investigation so imperfectly, that [his] results can lay little claim to accuracy. (Helmholtz, 1853 [1847], pp. 129–31; Smith, 1998, p. 135)

A second converse question concerned the opposite conversion of heat into work in heat engines. According to Helmholtz, "Whether by the development of mechanical force heat disappears, which would be a necessary postulate of the conservation of force, nobody has troubled himself to inquire" (Helmholtz, 1853 [1847], p. 135; Smith, 1998, p. 136), with the single exception of Joule's experiments comparing the expansion of compressed air either into the atmosphere or into a vacuum (see above).

In Germany Helmholtz's paper received a mixed but generally lukewarm reception. When it came to the attention of Thomson in 1852, however, it was enthusiastically embraced as providing support and systematic mathematical expression for the emerging "science of energy." Helmholtz was then quick to point out the similarities between his *vis viva* and "tension" and the newly defined "actual" and "potential" energies.

CONCLUSIONS: AN EXAMPLE OF SIMULTANEOUS DISCOVERY?

> The history of science offers no more striking instance of the phenomenon known as simultaneous discovery.... (Kuhn, 1977 [1959], p. 69)

During the second quarter of the nineteenth century, and particularly during the 1840s, it is possible—with the benefit of hindsight, at least—to identify at least a dozen authors who asserted important components of the principle that would be known as "the conservation of energy." A general belief in the correlation, conversion, and maybe fundamental unity and indestructibility of different natural "forces" was quite widespread among natural philosophers such as Faraday. The assertion, estimation, or measurement of an exact equivalence between heat and work was not infrequent, especially among writers like Joule from an engineering background. Within an essentially mechanical, Newtonian framework, Helmholtz had developed a principle of "the conservation of force." On the other hand, it is quite clear that these various insights had not crystallized into the coherent doctrine of the conservation of energy that was to emerge in the following decade (see chapter 5).

This may be partly accounted for by the various writers' mutual ignorance of each other's work, which can in turn be attributed to "the fact that so many of them wrote from different professional and intellectual backgrounds." However, "the men whom we call early exponents of energy conservation could occasionally read each other's works without quite recognizing that they were talking about the same things." Indeed, "no two of our men ever said the same thing. Until close to the end of the period of discovery, few of their papers have more than fragmentary resemblances retrievable in isolated sentences and paragraphs. What we see in their works is not really the simultaneous discovery of energy conservation. Rather it is the rapid and often disorderly emergence of the experimental and conceptual elements from which that theory was shortly to be compounded" (Kuhn, 1977 [1959], pp. 70–72). All the materials were on site, but the builders had yet to arrive.

At this point the final design of the new building was by no means obvious. Many were still cautious about the significance or even validity of Joule's findings. Joule himself became more concerned with establishing the dynamical theory of heat than any general correlation of forces. Helmholtz's account was arguably the most coherent and the closest to the later doctrines, but his vision did not flourish until it was transplanted from the realm of German physiology into the engineering-oriented culture of Britain's industrial north, with its emphasis on "work." But even in that context there were still important problems to be resolved, especially concerning the mismatch between the approaches of Joule and of Carnot.

5

ENERGY AND ENTROPY: THE BIRTH OF THERMODYNAMICS

INTRODUCTION

Not Copernicus and Galilei, when they abolished the Ptolemaic system;...not Newton,...not Young and Fresnel,...not Faraday and Clerk-Maxwell, in their splendid victory over *Actio in distans* [action at a distance]—more thoroughly shattered a malignant and dangerous heresy, than did Joule when he overthrew the baleful giant FORCE, and firmly established, by lawful means, the beneficent rule of the rightful monarch, ENERGY! (Anon., 1884)

1848, Year of Revolutions

Late in 1848, the Scottish businessman R. S. Newall—and Scottish businessmen were renowned as a hard-headed and, indeed, tight-fisted breed—offered the very substantial sum of £500 to "enable Regnault to complete his experiments on Steam and Effects of gases" (Lewis Gordon (1815–76), letter to William Thomson, quoted in Smith, 1998, p. 158). This clearly demonstrated the practical, commercial importance attached by Scottish engineers to Regnault's program of painstaking experimental measurement. More specifically, Newall was in partnership with Lewis Gordon, the professor of civil engineering and mechanics at Glasgow University, with the aim of commercial manufacture of a new, hopefully economical design of heat engine, the Stirling air engine. As Gordon joked in his letter to Thomson, playing on the engine's name and the common name for British currency, "the Sterling disinterestedness of the offer are manifest" (ibid).

Although Regnault had thus far been supported by the French government, his funding was threatened by the political turmoil of the time, when dissatisfaction with the restored monarchy had resulted in the eruption of a second republican revolution in Paris. In fact, 1848 came to be known as "the

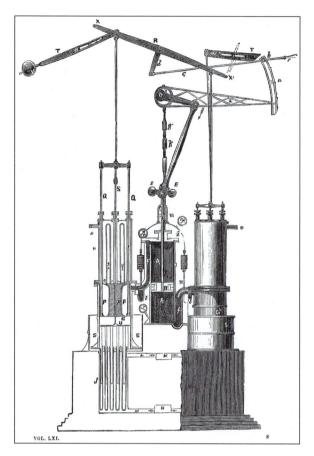

Figure 5.1: The hot-air engine patented by Macquorn Rankine and James Napier, on the cover of *Mechanics Magazine,* October 21, 1854; by permission of the Syndics of Cambridge University Library.

Year of Revolutions," marked by serious political unrest throughout Europe, with major upheavals in Italy, Austria, Germany, and, most notably, in the French capital. Ultimately, these disturbances seemed to bring to a climax and conclusion that period of profound social and economic change starting in the late eighteenth century that has been labeled "the Age of Revolution." The following decade ushered in a period of relative economic and social stability, symbolized by the first international trade fair, "The Great Exhibition of the Works and Industry of All Nations," held at the specially built Crystal Palace in London's Hyde Park in 1851. The fair was attended by some 6,039,195 visitors, many taking advantage of the railway network that had been constructed across Britain during the "railway mania" of the 1840s.

The same few years also witnessed the resolution of fundamental conflicts within the theory of heat, and the creation of a stable new theoretical order. In 1847 William Thomson and Joule finally met in the flesh at the BAAS meeting in Oxford, and in the same year Helmholtz's crucial paper on the "conservation of force" was published; 1848 and 1849 produced fruitful papers by Thomson on Carnot's theory of heat engines and on an absolute scale of temperature. Yet there remained serious theoretical contradictions between the ideas of Carnot and Joule, especially for Thomson. The way towards the resolution of these contradictions was first clearly shown in 1850 in a paper from the German academic physicist Rudolf Clausius. A steady stream of contributions followed over the next few years, and the emergent new theory of "the mechanical action of heat" was christened "thermodynamics" by Thomson in 1854. By 1857 an article by Thomson's friend and colleague J. W. Macquorn Rankine (1820–72) on "Heat, Theory of the Mechanical Action of, or Thermodynamics," in the popular *Cyclopaedia of the Physical Sciences,* spoke confidently of the new "SCIENCE OF THERMODYNAMICS"; the fundamental principles of the new science were codified as the First and the Second Laws of Thermodynamics, which were respectively based upon two distinct and radical new concepts, "energy" and what subsequently came to be known as "entropy."

Scientific Institutions in England

The new political stability from 1850 onwards was accompanied by reform to many of the institutions of science. The professionalization of science that had begun in France in the aftermath of the French Revolution, and been further promoted by early-nineteenth-century reforms to the German university system, had been slow to develop in England. It was only in 1833 that the very word "scientist" was coined by William Whewell (1794–1866), influential master of Trinity College, Cambridge, explicitly to label the new breed of increasingly professional, specialized, commercially oriented, and often secular experts in what was still widely known as "natural philosophy." But Whewell's term was slow to catch on and, more importantly, according to biologist T. H. Huxley (1825–95) writing in 1852 from a very practical point of view, scientific career prospects remained poor; it was possible for a scientist in England to "earn praise but not pudding." Since its foundation, the Royal Society, for example, had been partly a London club for gentlemen amateurs of science, some of whom were, of course, very expert in their fields—Joule and Darwin spring to mind—but many of whom were not. In the 1830s and 1840s the Society eventually reformed itself, and after 1847 admission to the Society was largely limited to men who were professional scientists of the new type.

The teaching of science at Oxford and Cambridge was also reformed following a Royal Commission of 1850–51. Since the mid-eighteenth century the "Mathematical Tripos" had been the core component of the Cambridge curriculum, conceived not as a professional training for mathematicians and scientists but as the best general intellectual training for future civil servants, colonial administrators, lawyers, and clergymen. Success in the arduous written examinations was highly competitive and socially prestigious; successful candidates were classified in ascending order as "Junior Optimes," "Senior Optimes," and "Wranglers," the top of the class being the "Senior Wrangler." The content of the course was "mathematical" in the traditional early modern sense: It included not just pure mathematics, but also applied or "mixed" mathematical disciplines. From the 1830s onwards, very much under the influence of Whewell, a line was drawn between such applied disciplines as astronomy, mechanics, and optics, which were considered to have arrived at a complete, stable, and thus "permanent" format, and such other topics as heat and electricity, which, although capable of mathematical expression, were still deemed to be "progressive" and unfinished. Only the former were considered to be suitable topics for mental training and examination. Despite this limitation, study for the Mathematical Tripos (examinations) provided the main vehicle for the teaching of science at the university. The 1851 reforms instituted a new "Natural Sciences Tripos" intended to provide a broader, less exclusively mathematical course in what was still commonly called "natural philosophy." Uptake for the new course was nevertheless slow until the 1880s, after the foundation of the new Cavendish Laboratory for experimental physics in the 1870s. The mathematicians for their part continued to regard experimental training as a dangerously time-consuming distraction from their purely intellectual discipline. (Of course, not everyone at Cambridge was scientifically inclined at all: As Maxwell warned Tait in 1873 as he was preparing to give a lecture to a distinguished academic audience at the University Senate House, "you will do very well, always remembering that to speak familiarly of a 2nd Law [of Thermodynamics] as a thing known for some

THE OLD, OLD STORY!

The Colonel. "Yes ; *He* was Senior Wrangler of his Year, and *She* took a Mathematical Scholarship at Girton ; and now they 're Engaged !"

Mrs. Jones. "Dear me, how interesting ! And oh, how different their Conversation must be from the insipid twaddle of Ordinary Lovers !"

THEIR CONVERSATION.

He. "And what would *Dovey* do, if Lovey were to die !"   *She.* "Oh, Dovey would die too !"

Figure 5.2: A cartoon from the humorous magazine *Punch* about a "senior wrangler," that is, the top mathematical physics student in his year at Cambridge University; Girton, the first college for women at Cambridge, had opened in 1873—although permitted to take the exams, women were not entitled to receive degrees until 1948. *Punch*, December 20, 1884, by permission of the Syndics of Cambridge University Library.

years, to men of culture who have never even heard of a 1st Law, may arouse sentiments unfavourable to patient attention" [quoted in Smith, 1998, p. 171]. This would probably be sound advice to the present day.)

THE RECONCILIATION OF CARNOT AND JOULE

The difficulty which weighed principally with me in not accepting the theory so ably supported by Mr. Joule was that the mechanical effect stated in Carnot's theory to be *absolutely lost* by conduction, is not accounted for in the dynamical theory otherwise than by asserting *it is not lost*. (From William Thomson's preliminary draft of 1853 [1851], published in Smith, 1976)

Thomson's Dilemma

William Thomson, knighted by Queen Victoria and then elevated to the peerage as Lord Kelvin of Largs in 1892—and often referred to by historians as "Kelvin" at any stage in his career—was arguably the dominant figure in British physical science in the second half of the nineteenth century. His output on a wide range of topics, but especially on heat and electricity and magnetism, was enormous. His interest and expertise in electricity led to his involvement in the laying in 1866 of the first successful transatlantic telegraph cable. Thomson was born in Belfast in Northern Ireland, but moved to Glasgow in 1832, when his father was appointed professor of mathematics at the university. In 1841 he went to Cambridge to study mathematics. After graduating in 1845 as "Second Wrangler" (that is, second from the top), Thomson went to Paris briefly to gain practical experience in Regnault's laboratory. He returned to Glasgow and was appointed professor of natural philosophy at Glasgow University in 1846, a post that he retained for over fifty years. Despite his education at the distinctively academic and Anglican University of Cambridge, Thomson remained deeply engaged with his Scottish heritage of engineering and Presbyterianism.

In 1849 Thomson was still torn between the competing claims of Joule and Carnot. On the one hand, Joule's experimental evidence for the conversion of mechanical work into heat was increasingly convincing, and his measurements of the mechanical value or "equivalence" of heat seemed to be increasingly reliable and accurate. On the other hand, Thomson was reluctant to abandon Carnot's elegant arguments and some of the consequences that stemmed from them, especially his own construction of an absolute temperature scale and the experimental confirmation of his brother James's prediction of the lowering of the melting point of ice. The conflict arose from the fact that, whereas Carnot's arguments were based upon the *conservation* of heat or caloric, Joule's whole program was aimed at demonstrating the *generation* of heat from mechanical work, and ultimately to proving that heat was just a form of motion.

At this time, Thomson still regarded heat as a function of state, so that the "quantity of heat" in a body in a given state must be fixed, and he believed that overall the total quantity of heat must be conserved, so that heat was not consumed in a heat engine. Thus Thomson was especially preoccupied by the *waste* of potential work (or mechanical effect) that seemed to occur when heat was simply *conducted* from a hot body to a cold one. He summarized the problem towards the end of his 1849 *Account of Carnot's theory:*

> When "thermal agency" is thus spent in conducting heat through a solid, what becomes of the mechanical effect which it might produce? Nothing can be lost in the operations of nature—no energy can be destroyed. What effect then is produced in place of the mechanical effect which is lost? A perfect theory of heat imperatively demands an answer to this question; yet no answer can be given in the present state of science. (Thomson, 1882–1911, 1 [1849], p. 118n; Smith, 1998, p. 94)

Thomson was not being unreasonable. In the 1790s measurements made by Watt's friend George Lee of the heat consumed in the furnace compared with the heat recovered in the condenser of a steam engine had revealed no detectable difference; in retrospect, this was understandable, since the efficiency of engines of the day was of the order of 2 percent. But even in Thomson's day serious scientific and commercial consideration was also given to a design by the Swede John Ericsson (1803–89) for an air engine that incorporated a "regenerator" intended to *recycle* the caloric supposed to power the engine, and thereby to operate with little or no fuel consumption; only in 1853 when a prototype vessel the *Ericsson* unfortunately sank during trials was the project abandoned. (Ericsson had greater success with his design for the "ironclad" warship USS *Monitor* in 1862.) Within this framework, the question of the disappearance of potential mechanical effect was very real.

Kelvin, Calvin, and Economy: Scottish Science in the Mid-Nineteenth Century

Lord, Thou hast made this world below the shadow of a dream,
An', taught by time, I tak' it so—excepting always Steam.
From coupler-flange to spindle-guide I see Thy Hand, O God –
Predestination in the stride o' yon connectin'-rod.
John Calvin might ha' forged the same—enormous, certain, slow –
Ay, wrought it in the furnace-flame—*my* "Institutio."
                    (Rudyard Kipling [1865–1936], *McAndrew's Hymn* [1893])

In the eighteenth century, Glasgow, on the banks of the river Clyde on Scotland's Atlantic coast, had emerged as one of the major ports of Britain's expanding trading empire and was indeed known as "the second city of the Empire," and Clydeside (Glasgow's docklands) became a key center for marine engineering. Some of the earliest steam-powered vessels had been built and operated on the Clyde at Glasgow; the *Comet* in 1812 was the first such vessel to be commercially successful in Europe. In the mid-nineteenth century, the transatlantic trade in particular offered a potentially lucrative opportunity for steam to compete with sail, and in 1850 the iron propeller-driven steamer, *The City of Glasgow,* began the first unsubsidized transatlantic passenger service.

The demand for reliability, robustness, and economy on such routes stimulated intensive development in marine steam engineering.

In contrast to the exclusively academic ethos of Cambridge, there existed in Glasgow close links between the commercial, engineering, and academic communities. The famous engineer James Watt had, of course, been instrument maker to Glasgow University, which in the nineteenth century boasted a chair of Civil Engineering and Mechanics, occupied by Lewis Gordon from 1840 until 1855, and thereafter by J. W. Macquorn Rankine (1820–72). Through the Glasgow Philosophical Society and commercially, the professors had close contacts with such ship and marine-engine builders as John Elder (1824–69) and James Robert Napier (1821–79). Thomson's family had close links with each community; his brother James Thomson (1822–92) trained as an engineer and took over the chair of engineering after Rankine's early death. Thus William Thomson, the mathematical physicist, was embedded in a network of practical and academic engineering connections that would hardly have been possible anywhere else in Britain.

Apart from their engineering interests, most of these men shared a devout Presbyterian background. Presbyterianism was (and is) a system of church organization based on the teachings of the sixteenth-century Reformation theologian John Calvin (1509–64), as expressed in his 1536 *Institutio [Institutes of the Christian Religion]*. Typically this involved the democratic choice of leaders within each congregation and sought religious understanding based on personal experience, in contrast and opposition to what was perceived as the authoritarian and obscurantist hierarchy of the Catholic Church. Although there were different currents within Presbyterianism, it usually involved the adoption of such distinctive Calvinist doctrines as the belief that both man *and* nature had been corrupted by the Fall. Man's only hope of salvation lay in a spontaneous gift of grace from God. Failure to accept God's gift was the cardinal sin, although Calvin's characteristic insistence upon the "predestination" of salvation was not always emphasized. The material world, corrupted by the Fall and thus containing the seeds of its own eventual destruction, was also perceived as a source of divine bounty for man's use, so here too the waste of resources was sinful. A preoccupation with economy, rooted or reflected in religious belief, was to find fruitful expression in commerce and engineering. This heritage is reflected in the Scots' reputation for thrift and, er, stinginess. The commercial virtue of such values are well represented by the cost-cutting tactics that enabled Scottish-born Andrew Carnegie (1835–1919), who emigrated to the United States in 1848, to dominate the U.S. steel industry in the late nineteenth century. The link with engineering is exquisitely evoked and examined in Kipling's "McAndrew's Hymn," the night-watch reminiscences of aging fleet engineer McAndrew—and in thermodynamics.

It has been argued—especially and with great cogency in Smith (1998)—that this concern for waste and efficiency also influenced the early articulation of thermodynamics. Certainly, the first-generation development of thermodynamics in Britain was dominated to an extraordinary degree by contributors who shared Thomson's Scottish and Presbyterian background. Apart from Rankine and the Thomson brothers, major contributions were made by Tait, the abrasive occupant of the chair of natural philosophy at Edinburgh, and by James Clerk Maxwell (1831–79), as well as by a number of lesser figures such as Balfour Stewart (1828–87) and the engineer Fleeming Jenkin (1833–85). With the addition of Joule from northerly Manchester, it is tempting to speak of a "North British School" in mid-Victorian energy science, representing a set of interests—religiously conservative yet commercially engaged—that was distinct both from the academic, Anglican mentality of Cambridge and from the liberal, professional, and religiously agnostic (or even atheist) "scientific naturalism" promoted by biologist Thomas Huxley (1825–95) and others in metropolitan London.

Clausius' Solution: Consumption *and* Transmission

That the competing claims of Carnot and Joule could be quite readily reconciled—in principle, at least—was first proposed by Clausius in a paper, "On the moving force of heat, and the laws regarding the nature of heat itself which are deducible therefrom," published in 1850. Clausius suggested that one could accept Joule's insistence on the precise equivalence of heat and mechanical work and also, at the same time and without inconsistency, affirm the "really essential portion" of Carnot's argument and thus most of his significant conclusions (Clausius, 1851 [1850], pp. 102–19; Smith, 1998, p. 99). This "really essential portion" was, according to Clausius, that a temperature difference and the transmission of heat from a hotter to a colder body were required for the production of work from any heat engine. Thus the key to reconciliation was in principle quite simple: "It is quite possible that in the production of work both [processes] may take place at the same time; a certain proportion of heat may be consumed, and a further portion transmitted from a warm body to a cold one; and both portions may stand in a certain definite relation to the quantity of work produced." (Clausius, 1851 [1850], pp. 3–4; Smith, 1998, p. 97). In other words, the *consumption* of heat, as proposed by Joule, and the *transmission* of heat as proposed by Carnot were both necessary to the generation of work. What had to be abandoned was Carnot's belief (at the time of writing the *Reflexions*) that the quantity of transmitted heat was conserved. Within a year this proposed new basis for the study of heat and work was endorsed by Thomson in a paper entitled "On the dynamical theory of heat" and thereafter rapidly elaborated in a stream of papers by Thomson, Rankine, and Clausius.

However, the reconciliation and integration of the theories of Carnot and Joule, as proposed by Clausius, although quite simple at first glance, proved extremely complex in its detailed development over the following decade, and far-reaching in its implications. There were two major consequences of the reconciliation. Firstly, Joule's equivalence of heat and work, soon codified as the First Law of Thermodynamics, was rapidly extended to the equivalence of all the various "forces of nature"; all physical phenomena—not just in heat and mechanics, but now including electricity and magnetism, light, gravity—were manifestations of the single new generalized concept of "energy," a universal currency of physical activity that was always and everywhere conserved. At the same time, the discipline of "physics" (and, indeed, science as a whole) came to be fundamentally restructured and unified around the "conservation of energy."

But neither the First Law nor the conservation of energy were sufficient foundation on their own for the science of thermodynamics. When Carnot was reformulated to fit with Joule, his logic was now seen to depend upon a new and essentially asymmetric or directional principle of natural behavior. At its crudest, this new principle, soon to be named the Second Law of Thermodynamics, affirmed that heat tends to flow from hot bodies to cold, and not the other way around—a kettle does not boil by sucking heat from the stove, although it could do so without infringing the First Law or the conservation of energy. This rough new principle was rapidly given precise mathematical formulation by Thomson and Clausius and was eventually crystallized by the latter into the concept of "entropy." The Second Law had some profound implications. Awareness of "time's arrow"—the fundamental difference between the past, the present, and the future in human experience—had always been a commonplace fact of life. Nevertheless, classical mechanics from Newton to Laplace was essentially reversible, or symmetrical with respect to time: Any sequence of collisions and interactions would ideally obey the same laws whether viewed forwards or backwards in time. But now the second major (and in some ways more novel) consequence of Clausius' synthesis was an explicit recognition and precise mathematical expression of the *directionality* (in time) of natural processes. The broader cosmological, metaphysical, and, indeed, spiritual implications of these ideas occasioned much debate about the ultimate fate of the universe and man's role within it.

FROM THE MECHANICAL EQUIVALENCE OF HEAT TO THE SCIENCE OF ENERGY

The very name "energy," though first used in its present sense by Dr. Thomas Young about the beginning of this century, has only come into use practically after the doctrine which defines it had...been raised from a mere principle of mathematical dynamics to the position it now holds of a principle pervading all nature and guiding the investigator in every field of science. (Thomson, 1881, p. 513; Smith, 1998, p. 8)

Equivalence, the First Law of Thermodynamics, and the Dynamical Theory of Heat

Clausius' synthesis of Joule and Carnot clearly required the wholehearted acceptance of the principle of equivalence of heat and work. As expressed by Clausius, "In all cases where work is produced by heat, a quantity of heat proportional to the work done is expended; and inversely, by the expenditure of a like quantity of work, the same amount of heat may be produced" (Clausius, 1851 [1850], p. 4; Smith, 1998, p. 98). Similarly, Thomson based his 1851 elaboration of Clausius on two propositions, the first of which was:

> PROP. I (Joule).—When equal quantities of mechanical effect are produced by any means whatever from purely thermal sources, or lost in purely thermal effects, equal quantities of heat are put out of existence or are generated. (Thomson, 1853 [1851], p. 264)

In 1857, in the first systematic exposition of the subject by Rankine, this became the "FIRST LAW OF THERMODYNAMICS.—*Heat and Motive Power are mutually convertible; and heat requires for its production, and produces by its disappearance, motive power in the proportion of 772 foot-pounds for each Fahrenheit Unit of Heat*" (Rankine, 1860 [1857], p. 413). In algebraic terms, work or motive power, W, and heat, Q, were connected by the formula $W = JQ$, where the conversion factor J depended upon the units in which W and Q were measured.

In general, of course, if one considers the heat and work inputs and outputs of a system, there are many circumstances in which heat may be added without the production of any obvious corresponding mechanical effect—in the melting of a solid, for example—or in which mechanical effect may be extracted without any thermal input—for instance in adiabatic expansion. A simple algebraic equivalence of the form $W = JQ$ was only directly applicable to a *closed* cycle of operations, such as Carnot's cycle. To provide a plausible explanation and accounting for any incomplete, noncyclic change, reference had to be made to some form of the dynamical theory of heat, that is, to the belief that the phenomena of heat could ultimately be explained in terms of the *motion* or *vis viva* of the constituent particles of matter. Joule's equivalence and the dynamical theory were mutually supportive: The dynamical theory offered an explanation of the alleged interconversion of heat and work, and the experimental demonstration of a precise and universal rate of exchange confirmed their identity. Early papers on the new synthesis were deeply engaged with dynamical theories. Clausius' first paper in 1850 devotes a considerable space to the explanation of "heat" in dynamical terms; Thomson's crucial 1851 paper was entitled, "On the dynamical theory of heat." An elaborate dynamical explanation of heat was offered by Rankine, who developed a theory of "molecular vortices," which attributed the expansive effect of heat to the centrifugal force of revolving particles of matter.

Whatever variety of the dynamical theory was adopted, a crucial concept was the "internal" energy, U, of a system. According to Clausius "latent heat," for example, was no longer to be explained in terms of heat somehow "hidden" as a caloric atmosphere among the particles of a liquid or vapor; rather the heat required for a change of state was *converted* into *work done* against intermolecular forces and maybe against the atmosphere in expanding a liquid into a vapor. Any difference between heat input and work output could now be attributed to a change in "internal energy"; in algebraic terms, $\Delta U = Q - W$, which is the form in which the First Law is now often presented.

Dynamical theories provided a general rationale for the identity and equivalence of heat and mechanical work and could resolve Thomson's persistent objection to Joule. Thomson, as we have already noted, had been particularly concerned with finding a convincing explanation for the "loss" of possible mechanical effect in the event of the straightforward conduction of heat from a hot to a cold body. A dynamical theory (as opposed to a "conservative" caloric theory) proposed that this "lost" mechanical effect could be located in the increased mechanical energies of the particles of the cold body where, although maybe *unavailable* for practical purposes, it had not been *destroyed* in any fundamental sense—although understanding of the *availability* of the mechanical effect remained a further, complex issue.

It is interesting to note that, despite the very rapid acceptance of the Clausius solution, direct *experimental* confirmation of the consumption of heat in a steam or other heat engine remained hard to come by; indeed, such measurements as had been made seemed to confirm the conservation of heat. Nevertheless, in the draft of his 1851 paper Thomson observed "that as yet no experiment can be quoted which directly demonstrates the disappearance of heat when mechanical effect is evolved; but [the author] considers that the fact has only to be tried to be established experimentally, having been convinced of the mutual convertibility of the agencies [i.e., heat and mechanical effect] by Mr. Joule's able arguments" (Smith, 1976, pp. 286–7; and 1998, p. 108). Experiments in the mid-1850s by the French engineer Gustave Adolph Hirn (1815–1890) did provide approximate confirmation of what already seemed to be widely regarded as a foregone conclusion.

Despite the crucial importance of the dynamical theory, most participants in the debate—especially those like Thomson broadly within the Cambridge school of mathematical physics—insisted on the independence of Joule's *experimentally established* equivalence, on the one hand, and the still varied and *speculative theories* about the ultimate nature of heat, on the other. Nevertheless, further development of a detailed, "microscopic" dynamical theory of heat was clearly desirable and received much attention in the following decades, firstly from Clausius himself and later from Maxwell and many others. However, the initial further development of the new "science of thermodynamics" can be treated separately from the development of the dynamical theory, which will be discussed in chapter 6.

From Equivalence to Energy

In the early 1850s Thomson and his associates, in particular Rankine, rapidly and vigorously promoted the extension of Joule's relatively narrow "mechanical equivalence of heat" into a radically new and general principle of "the conservation of energy." The generalized concept of "energy" and the axiomatic principle of its conservation provided a unifying foundation for the emergent discipline of modern physics. Thus some 20 years later, in the article on "Energy" in the *Encyclopaedia Britannica* (9th ed., 1875), William Garnett (1850–1932), Maxwell's assistant at the Cavendish Laboratory, asserted that "a complete account of our knowledge of energy and its transformations would require an exhaustive treatise on every branch of physical science, for natural philosophy is simply the science of energy." Although often presented by its protagonists as a fairly straightforward extension of classical Newtonian science, the new energy physics was in many respects profoundly different in both metaphysical foundations and in methodology.

The initial campaign to establish a physics of energy was waged in the early 1850s through a series of papers presented to various Scottish scientific societies and, most conspicuously, at the annual meetings of the BAAS. As early as 1854 in his address to the BAAS Thomson made the ambitious claim that Joule's discovery had "led to the greatest reform that physical science had experienced since the days of Newton" (Harman, 1982, p. 58). From the late-1850s onwards the program of the new "science of energetics" was formalized in a more systematic and popular mode through encyclopaedia articles and textbooks. Thomson and Tait's *Treatise on Natural Philosophy* (1867) was intended to provide the definitive reformulation of the foundations of physics.

The very rapid shift from a limited "equivalence" to universal "energy" can be clearly traced in these sources. In his fundamental 1851 paper, "On the dynamical theory of heat," Thomson presented Joule's "equivalence" as an instance of a more general "principle of mechanical effect"; this in turn was presented as a consequence of the conservation of *vis viva*, which was traced back to Newton. In 1852, writing after he had become acquainted with Helmholtz's previously neglected 1847 paper, Thomson spoke of the "conservation of mechanical energy" as an almost self-evident "universal truth" applicable to all physical phenomena:

> This state of certainty we may regard as now reached, in consequence of the recent advances of science in the establishment of the dynamical theory of heat and light.... *We may consequently regard it as certain that, neither by natural agencies of inanimate matter, not by the operations arbitrarily effected by animated Creatures, can there be any change produced in the amount of mechanical energy in the Universe;* and the belief that Creative Power alone can either call into existence or annihilate mechanical energy, enters the mind with perfect conviction. (draft of "Universal Tendency," quoted in Smith, 1998, p. 139; see also Smith and Wise, 1989, p. 384)

Rankine, in an 1853 paper, "On the general law of the transformation of energy," dropped the "mechanical," asserting confidently that "the law of the conservation of energy is already known, viz. that the sum of the actual and potential energies in the universe is unchangeable." By 1855 Rankine would read a paper that sketched the "outlines of the science of energetics."

Such generalization evidently required a redefinition of the very concept of "energy." The term had long been widely used in popular speech: According to the Romantic poet William Blake (1757–1827), in *The Marriage of Heaven and Hell,* "Energy is Eternal Delight." It was imported into mechanics as a synonym for *vis viva* by Thomas Young in 1807. But now its scientific application was to be extended far beyond such strictly mechanical boundaries. What *was* this new conserved entity? According to Rankine, writing in 1853, "the term *energy* is used to comprehend every affection [i.e., condition or state] of substance which constitutes or is commensurable with a power of producing change in opposition to resistance, and includes ordinary motion and mechanical power, chemical action, heat, light, electricity, magnetism, and all other powers, known and unknown, which are convertible or commensurable with these" (Rankine, 1881, p. 203; Smith,1998, p. 139).

Despite this all-embracing generalization of energy, it remained—for the time being, at least—grounded in mechanics. Mechanical energy, whether as the energy of motion or the energy of position, remained its privileged manifestation and measurement. In his encyclopaedia article, for example, in contrast to the foregoing comprehensive but vague "affections of substance," Rankine suddenly offered the much more prosaic definition "ENERGY, or the capacity to perform work" (Rankine, 1860 [1857], pp. 353–4; Smith, 1998, p. 165), resonating with the industrial, commercial origins of much of the new science—and a far cry from "eternal delight" for most industrial workers.

However defined, further elaboration of the concept was needed or, at least, some classification of its manifestations. On the one hand, as indicated by Rankine, there were various different forms of energy—such as mechanical, chemical, thermal, or electrical. On the other hand, there was a more fundamental—and controversial—distinction, that between actual and potential energy. The historical antecedents of the idea of the *conservation* of energy lay partly in eighteenth-century debate on the conservation of *vis viva,* but more immediately in a broader "conservation of mechanical energy" involved when—as in the swing of a pendulum, for example—energy of *position* was converted into energy of *motion,* and vice versa. An early attempt by Thomson to restructure physics in energy terms classified energy as "statical" or "dynamical"—vividly reflecting his mechanical starting point. Rankine, as already instanced in passing above, preferred the labels "actual" and "potential," or more broadly "actual or sensible" as opposed to "potential or latent." In their influential 1867 textbook *Treatise of Natural Philosophy,* Thomson and Tait opted for the now more familiar "kinetic" instead of "actual," somewhat to Rankine's annoyance.

For some scientists, however, the introduction of "potential" energy smacked of sleight of hand. According to the eminent astronomer Sir John Herschel (1792–1871), "the principle [of the conservation of energy]…is no other than the well known theorem of the conservation of *vis viva*…supplemented, to save the truth of its verbal enunciation, by the introduction of what is called 'potential energy'...[which] goes to substitute a truism for the announcement of a great dynamical fact" (Herschel, 1865, p. 439; Smith, 1998, p. 6). In other words, any apparent empirical imbalance in the conservation of energy could automatically and conveniently be attributed to some inaccessible "potential" reservoir.

Despite its clear mechanical pedigree, the new science of energy was not just an extension of Newtonian mechanics, a clarification of the "conversion of force" of Faraday and others; rather it represented a significant departure from the methods and assumptions of established physics. The dominant tradition in the first half of the nineteenth century remained the Laplacian (and ultimately Newtonian) program of explaining all phenomena in terms of forces acting between particles, often involving complex molecular models or hypotheses. The new science of energy aspired to replace this speculative, hypothetical component with an experimentally based, axiomatic system, and to substitute "energy" for "force" as the metaphysical foundation. In his introductory lectures at Glasgow in 1846 Thomson had maintained that physics was based on the laws of dynamics—physics was the science of force. From the early 1850s onwards physical science was to be based instead on the universal and indestructible foundation of "energy." Hence the somewhat overexcited reviewer of Joule's *Scientific Papers* in 1884 as quoted at the start of this chapter, exuberantly mixing theological, mythological, and political metaphors.

Thus the new physics was very far from being a simple refinement of the pre-1850 "conversion of force" promoted by Faraday. In fact, the basic concept of "energy," far from being Newtonian, is in some respects strangely archaic. The formless energy that persists through varied material manifestations is reminiscent of the *materia prima* (*first* or *fundamental matter*) of *Aristotelian* physics, the ultimate formless substance out of which all particular substances are formed. It is probably not accidental, therefore, that Rankine, who was widely read in the classics and philosophy, should have chosen the distinctively Aristotelian categories of "actual" and "potential" as the two basic forms of energy. Ironically, "energy" also owed a large debt to the "caloric," which it now finally and definitively replaced; turning again to Rankine's formulation, suggested alternatives to "actual" and "potential" were "sensible" and "latent," the main forms in which caloric had been supposed to exist. It could be argued that the idea of "latent" heat, banished by dynamical theories of phase change, reappeared as "potential" energy. According to Cannon (1978, p. 117), energy was "one single weightless fluid to replace the multiplicity favored in the 18th century."

Indeed, the concept of energy that emerged in the 1850s provided unity and coherence for a redefined discipline of "physics," fundamentally different from the traditional "natural philosophy"—despite the continued use of the latter term. The concept of energy not only unified the study of heat, mechanics, electromagnetism, light, and chemistry, it also allowed the new physics to claim ultimate authority over all other scientific discipline, from physiology to geology.

These changes were by no means purely conceptual and internal to science. Our retrospective focus has been mainly upon the often very abstract, conceptual development of early thermodynamics, apparently a matter of ideal gases and perfect heat engines. It is important not to underestimate, however, the importance of practical, experimental, and engineering interests in providing both incentive and material for theoretical discussion.

The works of Clausius and Thomson, in addition to their concern with ideal gases, both crucially involved a continuation of the interest, already shown by Carnot and Clapeyron, in the detailed behavior of steam and other vapors. Clausius, for example, reworked Clapeyron's discussion of latent heat and saturated vapor pressure to establish what is now known, reasonably enough, as the Clausius-Clapeyron equation. In combination with experimental determinations of the latent heats of different liquids and other data, this enabled estimates of the value of Carnot's function at different temperatures to be made, providing empirical support for the new synthesis. The full title of Thomson's 1851 paper is significant: "On the dynamical theory of heat; with numerical results deduced from Mr. Joule's 'Equivalent of a Thermal Unit' and Mr. Regnault's 'Observations on Steam.'" It shows "how the precise and extensive experimental data then being published by Regnault were providing a firm basis for the establishment of thermodynamics" (Cardwell, 1971, p. 240). Equally, the new concepts were rapidly applied to the improvement of power engineering; Rankine's 1859 *A Manual of the Steam Engine and Other Prime Movers* contained a lengthy chapter on the "Principles of Thermodynamics," which was later described by Maxwell as "the first published treatise on the subject." The new physics combined the mathematical rigor of Cambridge wranglers and German academics with the industrially driven experimental precision of Regnault and Joule and justified its national importance by its commercial applicability in manufacturing and transport.

Energy and Equity: A Digression on Physics and Capitalism

Energy corresponds to capital.

Doing work [upon a body] corresponds to buying.

Doing antiwork [i.e., work done by a body] corresponds to selling.

The transfer of capital is accompanied by two equal opposite acts, buying and selling, and it is impossible for one to go on without the other. Hence the algebraic sum of all the buying in the world is always zero: this is the law of conservation of capital. (Lodge, 1879, p. 283; Smith, 1998, pp. 291–2)

Just as the period from 1789 to 1848 has been labeled the Age of Revolution in European history, so the following period of great commercial investment and expansion from 1848 to about 1875 has been called the "Age of Capital." In the aftermath of the 1848 revolutions, the German economic and social theorist and activist Karl Heinrich Marx (1818–83) was expelled from Germany and settled in London, where he wrote a series of influential political and economic works, culminating in the monumental *Capital* in 1867. Thus, just as the economy and society were being restructured upon the basis of unified and interchangeable currency, so at the same time understanding of the natural economy was reformulated upon the basis of the universal currency of "energy."

The parallels were not lost on the scientists of the day. Popular scientific works in the later nineteenth century quite commonly developed an analogy with financial capital in order to explain energy—especially in its "actual" and "potential" forms. Balfour Stewart's 1878 *Conservation of Energy*, for example, contrasted energy of position and energy of motion: "The former may be compared to money in the bank, or capital, and the latter to money which we are in the act of spending...we can see the great capitalist, or the man who has acquired a lofty position, is respected because he has the disposal of a great quantity of energy...When the man of wealth pays the labouring man to work for him, he is converting so much of his energy of position into actual energy..." (Myers, 1985–86, p. 326). In his 1879 "Attempt at a Systematic Classification of the Various Forms of Energy" the British physicist (eventually Sir) Oliver Lodge (1851–1940) affirmed that energy "is power of doing work in precisely the same sense as capital is the power of buying goods" (Lodge, 1879, p. 279n; Smith, 1998, p. 291). The same analogy had already led Stewart and Tait, a couple of years after yet another French upheaval, the Paris Commune of 1870, to the intriguing proposition, in the context of irreversible change, that "the tendency of heat is towards equalization; heat is *par excellence* the communist of our universe, and it will no doubt ultimately bring the present system to an end" (Myers, 1985–86, p. 329)—although one doubts that the authors would have whole-heartedly embraced Marx's analysis of the internal contradictions of capitalism.

As Maxwell was to explain, the principle of the conservation of energy "indicates that in the study of any new phenomenon our first inquiry must be, how can this phenomenon be explained as a transformation of energy? What is the original form of the energy? What is its final form? And what are the conditions of transformation?" (Maxwell, 1876–7, p. 390 ; Smith, 1998, p. 126). The energy revolution replaced individualized interactions through forces acting on particles with accountable transfers and transformations of energy, comparing initial and final states—the bottom line, so to speak—rather than all the detailed transactions in between.

DISSIPATION, ENTROPY, AND THE SECOND LAW OF THERMODYNAMICS

Everything in the material world is progressive. (Thomson, preliminary draft of 1853 [1851]; see Smith, 1976, p. 282).

"The Really Essential Portion" of Carnot's Argument

According to Clausius, even when one had accepted the equivalence of heat and work and the inevitable destruction of a quantity of heat to produce work in any heat engine, it still remained possible to salvage "the really essential portion" of Carnot's argument, even though it had been based on the conservation of caloric. That essential portion, Clausius asserted in 1850, was that it was not heat as such but a *temperature difference* that was necessary to the operation of any heat engine. Over the following decade this insight was refined into various formulations of the Second Law of Thermodynamics.

However, the development of Carnot's argument had depended upon the further assumption that it was impossible to construct a perpetual motion machine. According to Carnot, a heat engine more efficient than a reversible engine could be operated back-to-back with a reversible engine, constantly producing a net output of work while the *conserved* heat or caloric circulated through the two engines: It would be a perpetual motion machine. But this argument would not be valid under the new principle that heat was being *converted* into work. Now the combination of a super-efficient engine with a reversible engine could transmit heat *from the cold reservoir to the hot*—without the input of any work, but nevertheless *without contravening Joule's equivalence of work and heat* or the principle of the conservation of energy. The heat thus pumped from cold to hot reservoir could then be used to generate useful work; such a device would thus constitute what came to be called a perpetual motion machine "of the second order": although it would not *create* energy out of nothing, it would, as Thomson explained, be "a self-acting machine [that] might be set to work and produce mechanical effect from the earth and the sea, or, in reality, from the whole material world" (Thomson, 1853 [1851], p. 265; Smith, 1998, p. 121).

Clearly an extra, new axiom was required upon which to base Carnot's arguments. For Clausius this was readily to hand in the observation that such behavior—the unaided transmission of heat from a cold to a hot body—contradicted the overwhelming human experience of heat, "which everywhere exhibits the tendency to annul differences of temperature, and therefore to pass from a *warmer* body to a colder *one*." This apparently banal observation, variously refined by Thomson, Clausius himself, and others, came to constitute what Rankine in 1857 labeled "The Second Law of Thermodynamics." For Thomson in 1851, for example, the basic axiom was that "it is impossible for any self-acting machine, unaided by any external agency, to convey heat from one body to another at a higher temperature" (Thomson, 1853 [1851], p. 265; Smith, 1998, p. 121). For Clausius in 1854 it was that "heat can never pass from a colder to a warmer body without some other change, connected therewith, occurring at the same time." These and other later formulations are recorded in the Sidebar.

It is worth noting that this new fundamental principle was not absolutely self-evident. In the case of a "burning glass," for example, that is a magnifying glass used to focus the rays of the sun to produce a spot of intense heat, it was by no means entirely clear that the temperatures produced were necessarily less than the temperature of the surface of the sun. If the spot were hotter, then heat would have been transferred from a cooler to a hotter state without any input work of work.

Despite its plausibility, the precise formulation and significance of the Second Law was much more problematic than the First Law had been. Of Rankine's formulation (see Sidebar: Variations on the Second Law), for example, Maxwell remarked that "its actual meaning is inscrutable" (Maxwell, 1878, p. 258; Smith, 1998, p. 165), while Tait in turn complained that "both Clausius & Rankine are about as obscure in their [thermodynamic] writings as anyone can well be" (1867 letter to Maxwell, quoted in Smith, 1998, p. 166). The Second Law was made more precise, but not necessarily more comprehensible, by the subsequent mathematical development of thermodynamics. Meanwhile, it turned out to have some unexpected consequences.

Variations on the Second Law

Clausius 1851 [1850]: Heat "everywhere exhibits the tendency to annul differences of temperature, and therefore to pass from a warmer body to a colder one."

Thomson, 1853 [1851]: "PROP.II. (Carnot and Clausius). – If an engine be such that, when it is worked backwards, the physical and mechanical agencies in every part of its motions are all reversed, it produces as much mechanical effect as can be produced by any thermo-dynamic engine, with the same temperatures of source and refrigerator, from a given quantity of heat."

Thomson, 1853 [1851]: "It is impossible, by means of inanimate material agency to derive mechanical effect from any portion of matter by cooling it below the temperature of the coldest of the surrounding objects."

Clausius, 1856 [1854]: "Heat can never pass from a colder to a warmer body without some other change, connected therewith, occurring at the same time."

Rankine, 1860 [1857] : "THE SECOND LAW OF THERMODYNAMICS. – If the total heat of a homogeneous and uniformly hot substance be conceived to be divided into any number of equal parts, the effects of those parts in causing work to be performed will be equal."

Clausius, 1867: "The entropy of the universe tends to a maximum."

Planck, 1932: "It is impossible to construct an engine which, working in a complete cycle, will produce no effect other than the raising of a weight and the cooling of a heat reservoir."

Carathéodory, 1909: "In the neighbourhood of any arbitrary initial state P_0 of a physical system there exist neighbouring states which are not accessible from P_0 along quasi-static adiabatic paths." (Constantin Caratheodory [1873–1950], a German-born mathematician of Greek family, "set himself the problem of finding a statement of the second law which, without the aid of Carnot engines and refrigerators, but only by mathematical deduction, would lead to the existence of an entropy function...." [Zemansky, 1957, pp. 172–3; see also Pippard, 1957, p. 30])

A Universal Tendency to Dissipation

One particular unforeseen and profound implication of the new Second Law was promptly articulated by Thomson. Heat behaved asymmetrically; there was an intrinsic grain to natural processes: Heat could readily flow from hot to cold, but not back again "against the grain." The new principle imposed a *direction in time* that had not been found in earlier physics.

In his first paper on thermodynamics the academic physicist Clausius had been mainly interested in *reversible* change as a definitive feature of Carnot's ideal heat engine. Thomson, on the other hand, with his vigorous, practical engineering interests (perhaps reinforced by a Calvinist abhorrence of waste), had always been interested in *irreversible* change, and especially in the potentially useful work that is lost or wasted in every real, less-than-ideal heat engine. Having endorsed the reconciliation of Joule and Carnot in his 1851 paper "On the dynamical theory of heat," in the following year in a paper, "On a universal tendency in nature to the dissipation of mechanical energy," Thomson developed the broader cosmological implications of the Carnot-Clausius theory, making public ideas that he had first explored privately in an unpublished draft version of the 1851 paper.

Thomson announced that the purpose of his 1852 paper was to "call attention to the remarkable consequences which follow from Carnot's proposition, established as it is on a new foundation, in the dynamical theory of heat." The Carnot-Clausius analysis of the heat engine implied that in any real heat engine—and thus in any natural thermal process—somewhat less than the ideal maximum "mechanical effect" would inevitably be extracted; an equivalent extra portion of heat would simply flow from the hot to the cold reservoir, and thus become unavailable as far as useful application was concerned. The extreme case would be the straightforward conduction of heat from a hot to a cooler body without the generation of any work whatsoever. According to Thomson, "there is an absolute waste of mechanical energy available to man, when heat is allowed to pass from one body to another at a lower temperature, by any means not fulfilling his [Carnot's] criterion of a "perfect thermodynamic engine"(Thomson 1852, 139–140; Smith, 1998, p. 124). The "absolute waste" followed from the fact that, although according to the new energy principles the heat has not been "annihilated," there was no way to recover and make use of it—other than by an additional input of work that would result in an equivalent waste somewhere else. As Thomson put it, "Any *restoration* of mechanical energy, without more than an equivalent of dissipation, is impossible in inanimate material processes" (Smith, 1998, p. 125).

From this scenario Thomson drew several truly "remarkable consequences." Thus far his account might have seemed to be of interest only to the professional heat engineer; Thomson's next step was to extend the analysis to a universal, global scale. Firstly and most fundamentally, he concluded that "there is at present in the material world a universal tendency to the dissipation of mechanical energy." Or, as he had expressed it somewhat more positively in his earlier unpublished meditations, "Everything in the material world is progressive. The material world could not come back to any previous state without a violation of the laws which have been manifested to man, that is, without a creative act or an act possessing similar power" (Smith, 1976, p. 282; Smith, 1998, pp. 110–11). Without, in other words, an act of God.

Newtonian mechanics, upon which physics had thus far been supposed to be founded, is essentially *symmetrical* in time; the gravitational interactions and motions of the planets, for example, should obey Newtonian dynamics whether viewed forwards or backwards in time. The Carnot-Clausius theory, on the other hand, and specifically the Second Law, seemed to introduce for the first time an intrinsic *direction in time* into the processes of the material world. The struggle to reconcile Newtonian mechanics and thermodynamics, and to understand the precise connection between the Second Law and time, continued through the remainder of the nineteenth century. Meanwhile, it emerged that these issues were not without cosmological and indeed theological significance for Thomson, as will be discussed at greater length below, after a brief account of the early mathematical development of the Second Law of Thermodynamics.

The Early Mathematical Development of Thermodynamics

The initial mathematical development of the implications of Clausius' reconciliation of Joule and Carnot was somewhat piecemeal. Broadly speaking, the application of Joule's equivalence to the behavior of gases confirmed some previous findings, and when applied to the Carnot cycle, it enabled the derivation of an expression for the long-sought Carnot function. At the same time Thomson constructed a new absolute thermodynamic scale of temperatures, which was the same as the standard ideal gas scale but shifted by $-273°$. In terms of this new scale, the Carnot function was simply equal to the absolute temperature. Finally there emerged a new state function, eventually named "entropy" by Clausius, which facilitated a concise, mathematical expression of the Second Law of Thermodynamics.

Thus the straightforward application of Joule's equivalence of heat and work to the behavior of ideal gases confirmed some earlier speculations about their specific heats. According to the dynamical theory, the heating (or cooling) of a gas compressed (or expanded) was no longer to be explained in terms of the squeezing out (or sucking up) of caloric, but was the result of the mechanical work done on (or by) the gas. Thus, provided that the internal energy of the gas remained constant, the heat required to expand a gas at constant temperature *must* now (according to Joule's equivalence) be equal to the work done by the gas in expanding. In his 1850 paper, therefore, Clausius was able to show that the difference between the two principal specific heats of ideal gases, $c_p - c_v$, was the same for all ideal gases and independent of temperature. Of course, Clausius was simply repeating what Joule (and Mayer and Holtzmann) had previously done as a means of determining a value for J. Indeed, the same conclusion had already been derived by Carnot from his own theory—that is, on efficiency grounds—and experimentally confirmed by Dulong in the late 1820s.

Even more interesting results were to be obtained by comparing deductions from Joule, on the one hand, and from Carnot and Clapeyron, on the other. Applying both approaches to the behavior of ideal gases, Clausius and Thomson each deduced an exact expression for Carnot's crucial function, $C(\theta)$ that determined the efficiency of heat engines; it turned out that

$$C(\theta) = (1/\alpha + \theta),$$

where $1/\alpha = 273$. This confirmed earlier empirical calculations that C increased with temperature. Published by Clausius in 1850 and by Thomson in 1851, such a formula had in fact been suggested by Joule in a private letter to Thomson as early as 1848.

In an important 1854 paper, Thomson reformulated matters in terms of a new absolute temperature scale. His earlier 1848 absolute temperature scale, it will be remembered, was *not* the same as the ideal gas scale. Now a new absolute scale was to be defined, such that "the absolute values of two temperatures are to one another in the proportion of the heat taken in to the heat rejected in a perfect thermo-dynamic engine working with a source and refrigerator at the higher and lower of the temperatures respectively" (Thomson, 1854). In other words, if heat Q_h was taken in at the higher temperature and heat Q_c given up to the cooler reservoir, then their absolute temperatures T_h and T_c were now given *by definition* by the formula

$$T_h / T_c = Q_h/Q_c.$$

It could easily be shown that in general absolute temperatures, $T = 1/\alpha + \theta = 273 + \theta$; the new absolute temperature scale was thus the same as the established ideal gas scale, except that the zero point was shifted to $-273°C$. Carnot's function now reduced quite simply to $C = T$, and the efficiency of an ideal Carnot engine would be just $\Delta T/T_h$.

At the same time, Thomson and Clausius both realized independently that the constraints on the conversion of heat into work in a simple reversible Carnot cycle operating between two reservoirs could be conveniently expressed in terms of the function Q/T. In his paper "On a modified form of the second fundamental theorem in the mechanical theory of heat," Clausius called the variable Q/T the "equivalence value of transformation." Some 10 years later, in 1865, he suggested a new name, "entropy," for this variable, which he designated by the letter S. The name, from the ancient Greek "trope" (τροπή) meaning "change," reflected its significance as an index of possible transformations and was conceived as a formal parallel to "energy," which is derived from the Greek "ergon" (έργον) meaning "work." Any more complicated reversible cycle could be broken down into a sequence of smaller cycles of absorption and emission at different temperatures. In these terms, throughout any such complex reversible cycle the sum, N, of all the equivalence value or entropy changes had to be zero.

Clausius went on to show that, for any *non*-reversible cycle, N must be *negative*, otherwise the cycle would be more efficient than Carnot's perfect reversible engine. From this it followed that, in any isolated system, spontaneous irreversible change must entail an *increase* in the entropy of that system. If equality to zero embodied the constraints upon a perfect thermodynamic engine (or the *impossibility* of extracting heat from a cold reservoir), "Clausius' inequality," as it came to be known, embodied in mathematical form the *inevitability* of waste, Thomson's "universal tendency to dissipation." Applying this analysis to the whole universe, envisaged as a closed system, Clausius arrived at the most economical expression of the fundamental principles of thermodynamics:

> We may express in the following manner the fundamental laws of the universe which correspond to the two fundamental theorems of the mechanical theory of heat.
>
> 1. The energy of the universe is constant.
> 2. The entropy of the universe tends to a maximum. (Clausius, 1867, p. 364–5)

Clausius' concept of entropy was initially received with considerable confusion, not to say hostility. The North British engineers were interested in efficiency and the "availability" of energy in a heat engine and interpreted Clausius in that light. In his *Sketch of Thermodynamics* (1868) Tait pronounced that "it is very desirable to have a word to express the *Availability* for work of the heat in a given magazine [reservoir]; a term for that possession, the waste of which is called *Dissipation*. Unfortunately, the excellent word *Entropy,* which Clausius has introduced in this connexion, is applied by him to the negative of the idea we most naturally wish to express.... And we take the liberty of using the term Entropy in this altered sense...." (Tait, 1868, p. 100; Smith, 1998, p. 257). It was left to Maxwell to point out delicately in a letter to Tait in 1876 that "when you wrote the *Sketch* your knowledge of Clausius was somewhat defective" (Smith, 1998, p. 257), and to confess that he himself, "under the conduct of Professor Willard Gibbs..." had been "led to recant an error I had imbibed from your [Tait's] *Thermodynamics* namely that the entropy of Clausius is *unavailable energy* while that of T[homson] is available energy.... The entropy of Clausius is neither the one nor the other it is only Rankine's Thermodynamic function" (Smith, 1998, p. 260).

Indeed, it remained unclear whether entropy corresponded to any real physical entity or was merely an artificial product of the mathematics. A satisfactory answer only emerged in the last quarter of the nineteenth century. Meanwhile, Thomson in particular pursued the more specific geological and cosmological implications of the universal tendency to dissipate energy, or to maximize entropy.

"The Earth Shall Wax Old": The Ages of the Sun and the Earth

> Of old hast thou laid the foundation of the earth: and the heavens are the work of thy hands. They shall perish, but thou shalt endure: yea, all of them

shall wax old like a garment; as a vesture shalt thou change them, and they shall be changed. But thou art the same, and thy years shall have no end.

—Psalm 102

In his 1852 paper Thomson proceeded from the very general universal tendency to dissipation to more specific conclusions about the origin, age, and fate of the Earth and the sun. Such reflections were by no means unprecedented; the ages of the sun and of the Earth had been matter for increasingly sophisticated discussion through the late eighteenth and early nineteenth centuries. Now, however, the Second Law appeared to categorically exclude any possibility of a steady-state scenario or process of cyclic regeneration. Since the heat of the Sun, whatever its source, must be finite, the constant, unrecoverable dissipation of its energy meant that its lifespan must also be finite. Similarly the Earth, constantly radiating more heat than it received, must once have been far hotter than at present, and would subsequently be far cooler. These arguments in general, and particular estimates of age, were employed by Thomson initially to attack so-called "uniformitarian" geology, and later—and somewhat ironically—against the progressive, evolutionary theory of Darwin.

There had, of course, been very vigorous discussion of the origins, age, and fate of the Earth, mainly in conventional geological terms, throughout the late eighteenth and early nineteenth centuries. The introduction of more specifically *physical* arguments was not entirely unprecedented. There had been two main issues connected with heat. On the one hand, there was the question of the cooling of the Earth, on the assumption of its molten origin, as proposed by the French naturalist Comte George-Louis Buffon (1707–88), for example. It was evident that heat from the sun was inadequate to maintain the present temperature of the Earth. In the absence of any recycling process, therefore, the Earth must be cooling: It must have previously been hotter and would subsequently be cooler. On the other hand, there was the question of the source—and thus the likely longevity—of the sun's energy. These topics had been discussed by Fourier in the 1820s, which had directly stimulated work by Thomson's Cambridge mathematics tutor William Hopkins (1793–1871) from the 1830s, and by Rankine in the 1840s. Also in the mid-1840s, Mayer developed a theory of meteoritic bombardment to explain the sun's heat. Thomson's inaugural address as professor of natural philosophy at Glasgow in 1846, was entitled "On the Distribution of Heat through the Body of the Earth." In fact, the main novelty in Thomson's 1852 presentation was the introduction of a strictly thermodynamic argument to oppose the possibility of any process of cyclic recharging of either the Earth's or the sun's resources of energy.

According to Thomson, given that the energy resources of the sun must ultimately be limited, it followed that "within a finite period of time past the earth must have been, and within a finite period of time to come the earth must again be, unfit for the habitation of man as at present constituted, unless operations have been, or are to be performed, which are impossible under the laws to which the known operations going on are subject" (Thomson, 1852,

p. 142; Smith, 1998, p. 125). This dry, published formulation obscures the theological resonance revealed in his earlier draft contemplations: "I believe the tendency in the material world is for motion to become diffused, and that as a whole the reverse of concentration is gradually going on—I believe that no physical action can ever restore the heat emitted from the sun, and that this source is not inexhaustible; ... The earth shall wax old, &c. The permanence of the present forms & circumstances of the physical world is limited" (Thomson, 1851, in Smith, 1976, p. 312). The new science of thermodynamics, therefore, appeared to establish the transitory nature of the material world—certainly as far as human life within the solar system was concerned—in keeping with a liberal interpretation of the Bible and, more specifically, in keeping with a Calvinist view of a "fallen" nature containing the seeds of its own destruction.

The main target—of many of the British speculations, at least—was the "uniformitarian" or "non-progressive" geology of (Sir) Charles Lyell (1797–1875), which claimed that conditions on Earth were essentially stable and static, and which was vigorously opposed by the Anglican tradition of geology dominant in Oxford and Cambridge. Thus Hopkins, at the conclusion of a lengthy exposition in his presidential address to the BAAS in 1854, asserted that "the heat of the sun must ultimately be diminished, and the physical condition of the earth therefore altered, in a degree inconsistent with the theory of non-progression." With limited and constantly dissipated resources, the Earth similarly must have had hot and *finite* past—and a cold and uninhabitable future. In an 1854 paper Thomson estimated the present age of the sun as about 32,000 years and its further life-span as some 300,000 years. By the early 1860s, having relinquished Mayer's meteoric theory in favor of a theory developed by Helmholtz postulating the release of energy through the sun's gradual gravitational collapse, Thomson increased his figure for the lifespan of the sun to between 50 and 500 million years. Reassuringly, he arrived at similar figures derived from the cooling of the Earth.

Ironically, in the 1860s, after the publication of *On the Origin of Species*, the religiously conservative Thomson would find himself trying to undermine Darwin's "progressive" theory of evolution on the grounds that the estimated age of the Earth was insufficiently long to allow for gradual "evolution by natural selection." While most geologists were initially prepared to accept the authority of the physical arguments for the above figures, when Thomson subsequently reduced his best estimate to 24 million years there was widespread rejection. Only with the discovery of radioactivity and its heating effect at the end of the century was the conflict resolved.

Thomson's extension of the scope of thermodynamics from engineering to cosmology was original and unexpected. Helmholtz observed that "we must admire the sagacity of Thomson, who, in the letters of a long-known little mathematical formula which speaks only of the heat, volume, and pressure of bodies, was able to discern consequences which threatened the universe, though certainly after an infinite period of time, with eternal death" (Helmholtz, 1856 [1854], p. 503; Smith, 1998, p. 142). The prompt further extension of Thomson's "remarkable consequences" from the fate of the solar system to the fate of the cosmos at large had been made by Rankine at the 1852 meeting of the BAAS; Thomson's "universal tendency," he suggested, implied "the conversion of all the other forms of physical energy into heat, and to the equable diffusion of all heat; a tendency which seems to lead towards the cessation of all phenomena" (Rankine, 1852; Smith,1998, p. 142). As the title of Rankine's talk, "On the Reconcentration of the Mechanical Energy of the Universe," indicates, however, he wondered whether "the world, as now created, may possibly be provided within itself with the means of reconcentrating its physical energies, and renewing its activity and life." Perhaps diffused radiation would be reflected at the boundaries of a possibly finite interstellar aether, and re-focused at points within it? On the whole, however, Thomson's basic conclusion was widely accepted. Clausius concluded that, when the entropy of the universe had finally reached a maximum, no further change would be possible and "the universe would be in a state of unchanging death." This gloomy message was enthusiastically reworked towards the end of the century.

CONCLUSION

The 1850s witnessed a profound change in scientists' understanding of heat in particular and of the physical sciences in general. By the 1860s, acceptance of Joule's "mechanical equivalence of heat" had resulted in the final demise of the material, caloric theory in favor of the *dynamical theory of heat*. The combination of Joule's insights with Carnot's analysis of the heat engine had led to the recognition of a *directionality*, a "universal tendency to dissipation," in all natural heat processes. More generally, physics as a discipline had been restructured on the basis of the universal currency of *energy*. But developments in thermodynamics were by no means complete, nor was there unanimity about their significance. The precise significances of the Second Law, dissipation, and entropy remained a matter of vigorous debate throughout the remainder of the nineteenth century and beyond. In particular, there was disagreement over the relationship between thermodynamics as an empirically grounded but purely descriptive mathematical framework, on the one hand, and speculative attempts to provide mechanical, molecular explanations of the detailed thermal properties of matter, on the other. These debates took place pre-eminently within the context of the dynamical theory of heat and the "kinetic theory" of gases, which Clausius, Maxwell, and others developed from the mid-1850s, as is described in the following chapter.

6

THE KIND OF MOTION WE CALL HEAT: THE DEVELOPMENT OF THE KINETIC THEORY OF GASES

INTRODUCTION: FROM THE DYNAMICAL THEORY TO STATISTICAL MECHANICS

[W]e must look steadfastly into this theory which calls heat a motion. (Clausius, 1851 [1850], pp. 3–4; Smith, 1998, p. 97)

Brownian Motion and the Kinetic Theory

In an "Autobiographical Chapter" written late in life, Charles Darwin (1809–1882) recollected that, as a young man, before his career-defining voyage around the world on HMS Beagle (1831–36),

> I saw a good deal of Robert Brown, "facile Princeps Botanicorum," as he was called by Humboldt. He seemed to me to be chiefly remarkable for the minuteness of his observations and their perfect accuracy...I called on him two or three times before the voyage on the *Beagle*, and on one occasion he asked me to look through a microscope and describe what I saw. This I did, and believe now that it was the marvellous currents of protoplasm in some vegetable cell. I then asked him what I had seen; but he answered me, "That is my little secret." (Darwin, 1892, p. 46) (Alexander von Humboldt [1769–1859], the most famous naturalist of his day, had inspired Darwin's voyage.)

The enigmatic Robert Brown (1773–1858), glowingly endorsed as "easily the foremost of botanists," is chiefly remembered today as the discoverer of "Brownian motion": viewed through a microscope, very small particles suspended in fluid are often seen to be in a state of incessant, random agitation. Maybe this was what Darwin saw. In modern textbooks this phenomenon is often presented as a vivid confirmation of the ceaseless, random motion that, according to "kinetic theory," is characteristic of the invisible, submicroscopic molecules that make up liquids and gases.

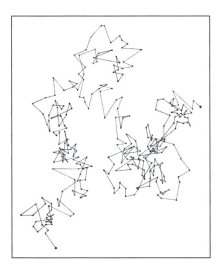

Figure 6.1: The random "Brownian" movement of a microscopic grain of mastic suspended in water, charted at 2-minute intervals. Jean Perrin, *Atoms* (London, 1923); Whipple Library, University of Cambridge.

Very little notice was taken of Brownian motion, however, during the rapid development of modern kinetic theory from the late 1850s onwards. Although the theory was very successful in explaining many properties of gases, it was subject to renewed skepticism towards the end of the century, and the physical reality of atoms and molecules was seriously challenged. It was at this juncture that Brownian motion finally played a crucial role, and the kinetic theory was spectacularly vindicated in the first decade of the twentieth century, when Albert Einstein's (1879–1955) kinetic-theory-based calculations of the random motions of Brownian particles were confirmed experimentally.

Overview

In the aftermath of the construction of thermodynamics and the science of energy, "heat," in a sense, ceased to exist. The old kingdom of heat was overwhelmed by the new empire of energy The range of phenomena that had once without much problem been recognized to fall within heat's domain was now broken up and parceled out among other disciplines. The phenomena of "radiant" heat, for example, were increasingly linked to light and the new science of electromagnetism; chemistry, which had largely neglected heat issues in the first half of the nineteenth century, took a renewed interest in the dynamics—as opposed to the reagents and products—of chemical reaction, resulting in the development of chemical thermodynamics towards the end of the century. The bulk of heat's domain, however, was taken over by mechanics, initially in the form of thermodynamics, and then increasingly towards the end of the century in the form of "statistical mechanics." It is with this mechanical development that we shall be largely concerned in the present chapter.

The acceptance of Joule's "mechanical equivalent of heat" in 1850 was widely taken to imply the acceptance in principle of the "dynamical" theory of heat, the theory that heat was fundamentally "motion." As we have already seen, a variety of dynamical theories had been proposed in preceding decades, varying in their notion of what kind of motion was involved—be it translation, rotation, or vibration—and of what exactly was moving—be it atoms and molecules, their "atmospheres," or a combination of the two. Yet in the early 1850s, apart from Rankine's theory of "molecular vortices," there were no immediate further developments: In fact, much emphasis was laid on the *independence* of the laws of thermodynamics and energy from any speculative, mechanical hypotheses about the basic structures and behavior of matter.

This caution was abandoned in the later 1850s. In the following two decades Clausius, Maxwell, and others began a vigorous development of mechanical, atomic models of matter; attention concentrated (once again) on gases, resulting in "the kinetic theory of gases," as it was subsequently labeled by Maxwell (in his 1871 *Theory of Heat*). This provided successful explanations of many properties of gases, both in general qualitative terms (e.g., diffusion) and in precise quantitative terms (e.g., viscosity), and the theory was widely accepted by the 1870s.

From a theoretical point of view, one of the most remarkable features of this development was the introduction by Maxwell of *probabilistic* and *statistical* methods for describing the motions of large collections of atoms and molecules, recognized as in some senses random. Despite the success of these methods, there remained an underlying tension between those (mainly German) physicists who believed that the descriptive laws of thermodynamics could be reduced to mechanics, and those physicists (especially North British) who believed in the independence and irreducibility of thermodynamics, especially the Second Law. This tension, combined with some perceived inconsistencies and failings of the theory, seems to have contributed to that reaction against it towards the end of the century that was largely resolved by Einstein in 1907.

REVIVAL AND DEVELOPMENT OF THE KINETIC THEORY OF GASES, 1856 TO 1870S

In reading Clausius, we seem to be reading mechanics; in reading Maxwell, and in much of Boltzmann's most valuable work, we seem rather to be reading in the theory of probabilities. (Gibbs, 1948, 2, part 2, p. 265 [1889]; Brush, 1976, p. 170)

Elements of the Kinetic Theory

Thanks to Newton, from the early eighteenth through the first half of the nineteenth century, a gas was commonly supposed to consist of a more-or-less static lattice of atoms or molecules, relatively widely spaced compared to those in a solid or liquid, of course. Even most early nineteenth-century *dynamical* (as opposed to caloric) theories of heat imagined gases in this way, heat being attributed to the energy of the rotations and/or vibrations of the molecules and/or their atmospheres. The *kinetic* theory of gases imagined instead that a gas consisted of a vast number of widely separated particles or molecules that, far from occupying fixed positions, were in constant rapid motion (translation) through space; these particles were imagined to travel randomly in all directions in straight lines, their paths diverted only by collisions with each other or with the walls of any container. The pressure of a gas was supposed to be the result of the continual bombardment of the walls of any container by these fast-moving particles, and heat and temperature were connected to the kinetic energy of the particles. Initially this was distinctively a theory of gases, for it was assumed that in solids and liquids the constituent particles were much more tightly packed together and thus unable to travel any great distance without collision. Subsequent developments, theoretical and experimental, made the distinction between gases and liquids, at least, far less clear cut.

The kinetic theory had two major strengths. Firstly, in general qualitative terms it provided a plausible model of such processes as gaseous diffusion (the way in which one gas will mix with or diffuse into another) or the process of evaporation from a liquid (fast-moving molecules in the liquid could escape from its surface). It is worth recalling Dalton's struggle to construct a convincing model of the mixing of gases according to the static theory—although this did have the side effect of stimulating him to develop his atomic theory. Secondly, in precise quantitative terms, it was possible to derive Boyle's law from the kinetic hypothesis. In 1738 Daniel Bernoulli had shown that $PV = 1/3\ Nmv^2$, which would be constant if the velocity, v, remained constant. In 1845 Waterston had made more explicit the further implication that temperature was proportional to the kinetic energy of the particles. Against this, however, Newton's hypothesis of a mutual inverse-linear repulsion between gas molecules also gave Boyle's law, and carried the additional authority of Newton's mighty name. In the early 1850s, therefore, apart from Waterston's unfairly neglected contributions, the kinetic theory still remained peripheral and relatively undeveloped.

Revival of the Kinetic Theory: Krönig and Clausius

Overt interest in the kinetic theory of gases was re-awoken in 1856 when the German chemist August Karl Krönig (1822–79) published a paper on the "Outlines of a theory of gases" which developed a rudimentary kinetic theory of gases. Although of no great theoretical originality, this paper prompted Clausius to publish his own ideas on the subject the following year, in a paper provocatively entitled "On the nature of the motion that we call heat." His theory was immediately more complex than Krönig's, as he explained:

> Krönig assumes that the molecules of a gas do not oscillate about definite positions of equilibrium, but that they move with constant velocity in straight lines until they strike against other molecules, or against some surface which is to them impermeable. I share this view completely, and I also believe that the expansive force [i.e., the pressure] of the gas arises from this motion. On the other hand, I am of the opinion that this is not the only motion present.
>
> In the first place, the hypothesis of a rotary as well as a progressive [translational] motion of the molecules at once suggests itself; for at every impact of two bodies, unless the same happens to be central and rectilineal, a rotary as well as a translatory motion ensues.

I am also of the opinion that vibrations take place within the several masses [molecules] in a state of progressive motion. Such vibrations are conceivable in several ways. Even if we limit ourselves to the consideration of the atomic masses solely, and regard these as absolutely rigid, it is still possible that a molecule, which consists of several atoms, may not also constitute an absolutely rigid mass, but that within it the several atoms are to a certain extent moveable, and thus capable of oscillating with respect to each other. (Clausius, 1857; Brush, 1976, p. 172)

And Clausius' development of the theory was more detailed and original. Whereas Krönig had assumed that at a given temperature all molecules moved at the same speed (and even parallel to a set of Cartesian axes), Clausius recognized that "we must assume that the velocities of the several molecules deviate within wide limits on both sides of the average value" (Clausius, 1857; Brush, 1976, p. 174); nevertheless, for the most part he contented himself with using just average values. He readily deduced the Boyle's law equation $PV = constant = 1/3\ Nmv^2$, and comparison with the ideal-gas equation, $PV = RT$, suggested that temperature was proportional to the kinetic energy of the gas molecules. His more complex model of molecular motions, however, made Clausius the first to insist that it was solely the *translational* kinetic energy that determined the temperature and allowed him to develop a more refined account of the specific heats of gases, which will be discussed in the next section.

Clausius also discussed the general behavior of gases and liquids. As noted above, gaseous diffusion was especially well explained by the kinetic theory: Two bodies of gas brought into contact would evidently rapidly mingle. He also gave an account of evaporation: Given that there would be some variation in the velocities of the molecules in a liquid, even below its boiling point a proportion of molecules would have enough velocity to escape from the body of the liquid. Moreover it was possible to make precise quantitative predictions. From the basic Bernoulli equation it was simple to calculate—and Herapath, Waterston, and Joule had already done this—the average velocities at which gas molecules must be moving; for oxygen and hydrogen, for example, Clausius derived the values 461 and 1,844 m/s respectively.

There was a problem, however. Although these figures tallied reasonably well with the speed of sound in different gases, it was pointed out that they seemed to imply that gaseous diffusion should also be a very fast process, whereas everyday experience in the chemistry laboratory—and indeed the wider world—indicated that it was relatively slow. Clausius promptly tried to explain this contradiction in 1858 in a paper, "On the mean lengths of the paths that are covered by individual molecules...." He emphasized the point that, in a real gas, the molecules would not travel in uninterrupted straight lines, but they would be constantly colliding with each other, thereby changing speed and direction at random. The overall diffusion of one gas into another would thus take place much more slowly than the motion of the individual molecules. Clausius went on to derive an expression for the average distance travelled between collisions—or the "mean free path," as it is now known—of a molecule in terms of the size of the molecule and the overall density of the gas. Unfortunately, at that time reliable estimates of the sizes of individual molecules were not available—indeed their very reality was still in question; rough figures seemed to suggest, however, that the mean free path would be short enough to slow down diffusion to observed rates.

Maxwell, Boltzmann, and Statistical Methods

At this point Clausius' speculations came to the attention of the brilliant young Scottish physicist James Clerk Maxwell. Maxwell shared Thomson's Scottish heritage, but came from a gentlemanly farming family rather than an urban commercial background. Educated at Cambridge, he graduated Second Wrangler in 1854, despite some misgivings about the value of his studies:

> In the grate the flickering embers
> Served to show how dull November's
> Fogs had stamped my torpid members,
>   Like a plucked and skinny goose.
> And as I prepared for bed, I
> Asked myself with voice unsteady,
> If of all the stuff I read, I
>   Ever made the slightest use.
>
> ("A Vision of a Wrangler, of a University, of Pedantry,
> and of Philosophy," Nov. 10, 1852. In Campbell
> and Garnett, 1882, p. 612; Smith, 1998, p. 214)

He nevertheless went on to become professor of natural philosophy at Aberdeen and then in London, before being appointed the first head of Cavendish Laboratory for experimental physics when it finally opened in Cambridge in 1874. Maxwell is probably best known for his development of an integrated mathematical theory of electricity and magnetism, on the basis of which he predicted that light (and radiant heat) consisted of electromagnetic waves. But he also made a profound contribution to the development of kinetic theory, which will be discussed below. His work was sadly curtailed by his premature death in 1879.

In response to Clausius' speculations, Maxwell published in 1860 "Illustrations of the dynamical theory of gases" and in 1867, partly in response to criticisms by Clausius, a considerably revised account entitled "On the dynamical theory of gases." The distinctive feature of Maxwell's theorizing was his use of *statistical* methods. As we have seen, Clausius was well aware that there might be considerable variation in the speeds of molecules, and his derivation of the mean free path was partly based on probabilistic considerations; nevertheless, he preferred whenever possible to deal simply with the average values of variables. Maxwell's approach, in contrast, was thoroughly statistical from the very start, proposing that the distribution of the velocities of molecules followed the same "normal" bell-shaped distribution as experimental errors. This meant that the speeds of the molecules in a gas would cluster around the average value, but that a certain precisely defined proportion of higher and lower speeds would also be found. Maxwell's formula also showed how the distribution of speeds would change with temperature. More-or-less direct experimental confirmation of Maxwell's distribution would eventually be obtained, but only some 60 years later.

On this statistical basis Maxwell developed a detailed mathematical treatment of various "transport properties" of gases, notably diffusion, viscosity— that is, "stickiness," or the internal resistance that is offered to motion; e.g., oil is more "viscous" than water—and heat conduction. Initially he used the mean-free-path concept, but in the later paper Maxwell switched to an analysis based directly upon molecular collisions, an approach that provides the foundation for the modern treatment of the subject. One prediction of his theory was that many transport properties of gases—and particularly the viscosity—should be independent of the density (or pressure) of the gas (other things being equal). At first sight it seemed most improbable that the viscosity of a gas should stay the same as its density (or pressure) was increased or decreased. As Maxwell remarked, "This is certainly very unexpected, that the friction should be as great in a rare as in a dense gas" (1860; Brush, 1976, p. 190). Although very little work had been done on the subject, such experiments as had been done seemed to show that viscosity was *not* independent of pressure. Accordingly Maxwell, in cooperation with his wife Katharine, conducted further experiments on the viscosity of air; they found that the viscosity did indeed remain constant as the pressure was varied between 1/2 and 30 inches of mercury (1/60 to 1 atmosphere). According to John William Strutt, Lord Rayleigh (1842–1919), who succeeded Maxwell as head of the Cavendish, "in the whole range of science there is no more beautiful or telling discovery than that gaseous viscosity is the same at all densities" (Rayleigh, 1890; Brush, 1976, p. 193).

The publication of Maxwell's theories stimulated much of the further work on the kinetic theory in the later nineteenth century, most notably by the Austrian physicist Ludwig Boltzmann (1844–1906). Boltzmann vigorously developed Maxwell's statistical analysis of systems of colliding particles. An early paper in 1868 incorporated the effect of an external force (such as gravity) on the distribution of molecular velocities in a body of gas. The resultant extra "Boltzmann factor" enabled calculation of the distribution of density and temperature in a vertical column of gas, a scenario that others, most notably Waterston, had previously investigated because of its relevance to the structure of the Earth's atmosphere. The same analysis turned out to be applicable to the distribution of macroscopic particles suspended in a liquid and was exploited to good effect by Einstein in his treatment of Brownian motion.

In the absence of the fairly direct experimental confirmation that became available later, Maxwell's proposed velocity distribution, though plausible and backed by the experimental confirmation of some of its predictions, nevertheless remained somewhat arbitrary. In 1872, therefore, in a paper entitled "Further studies of thermal equilibrium among gas molecules," Boltzmann

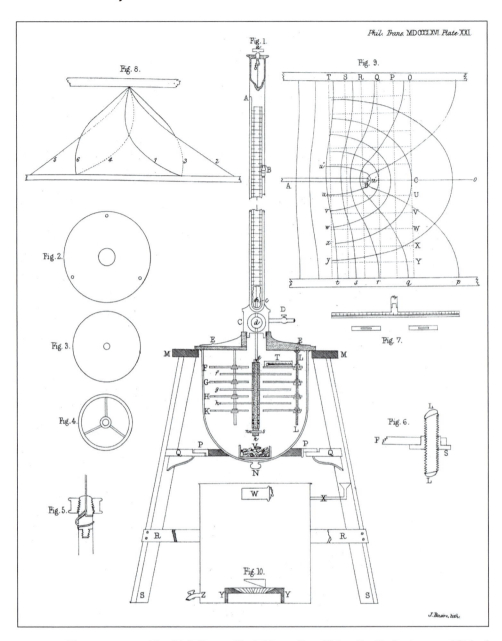

Figure 6.2: The apparatus with which James Clerk Maxwell and his wife, Katharine, established that the viscosity of a gas is, surprisingly, independent of its pressure. "On the Viscosity or Internal Friction of Air and other Gases," *Philosophical Transactions of the Royal Society* (1866); by permission of the Syndics of Cambridge University Library.

attempted to establish it on more secure, mechanical foundations. He constructed a new function of molecular velocities—the "H-function," roughly the negative of entropy—and showed that, for any distribution of velocities other than the Maxwellian, this function would tend to decrease as a result of molecular collisions; in other words, any distribution other than Maxwellian would inevitably change to Maxwellian, which alone was stable.

Shortly after the publication of this paper, Maxwell remarked dryly upon their very different scientific styles: "By the study of Boltzmann I have become unable to understand him. He could not understand me on account of my shortness and his length was and is an equal stumbling block to me. I am very much inclined to join the glorious company of supplanters [*usurpers?*] and to put the whole business in about six lines" (Maxwell to Tait, 1873, in Maxwell 1990–2002, 2, p. 915–6; Smith, 1998, p. 263). (This is very reassuring for anyone who has struggled to understand either Boltzmann or Maxwell—or, indeed, both.) However, their differences were not just stylistic. The H-function approach seemed to Boltzmann to provide also a rigorous mechanical explanation of the Second Law of Thermodynamics, and thus of the inevitable increase of entropy in irreversible processes and in the cosmos as a whole. In seeking thus to reduce the Second Law to mechanics, Boltzmann was in fundamental disagreement with Maxwell (and many other North British physicists), who saw the Second Law as an essentially independent—and profoundly moral—component of thermodynamics. The ensuing debate will be discussed in more detail below. Meanwhile, there were more specific challenges for the kinetic theory to deal with.

PROBLEMS, REFINEMENTS, AND RECEPTION

Considering the number and variety of the phenomena of gases which are accounted for on this [dynamical] theory, and especially the exactness with which it accounts for the hitherto inexplicable phenomena of diffusion, there seems to be a considerable probability in its favour. The small discrepancy between the observed and computed ratios of the specific heats (1.42 and 1.33) may be found to proceed from some property of the particles not taken account of in the mathematical analysis. (Simon Newcomb, 1862; Brush, 1976, p. 203)

Despite Maxwell's striking success with viscosity, however, some predictions of the kinetic theory did not square so easily with experiment. Well-established values of the specific heats of gases, in particular, proved persistently difficult to explain in a satisfying manner. Thus there was clearly considerable scope for further refinement of the theory, and considerable further progress *was* made in developing more complex "equations of state" that would explain some of the aberrations from ideal-gas behavior observed in real gases. Overall, from the mid-1860s onwards the kinetic theory was widely accepted as, at the very least, a very plausible model of the structure of gases.

The Problem of Specific Heats

The basic kinetic theory had led to the proposition that the temperature of a gas was directly proportional to the average *translational* kinetic energy of the gas molecules. This connection between energy and temperature allowed an estimation of the specific heat, c_v. In the case of a simple monatomic gas— when all the internal energy of the gas would be in the form of translational kinetic energy—a value was derived, which meant that γ, the ratio of the two specific heats c_p / c_v, would be 5/3 or about 1.67. Unfortunately, the monatomic inert gases (helium, neon, and so on) were not isolated until the end of the century, but experiments conducted in 1875 gave a figure very close to this value for mercury vapor; if mercury vapor were monatomic, then this seemed to be good confirmation for the kinetic theory.

This approach was less successful when applied to more complex molecules made up of two or more atoms. As noted above, Clausius insisted that more complex molecules could well have rotational and vibrational motions as well as just translational motion. If temperature depended solely on the translational kinetic energy, then extra energy would nevertheless be needed for any given temperature increase in order to increase the rotational and vibrational components of motion; thus gases formed by complex molecules should have higher c_v than single atoms.

But how much extra energy should be needed? How would the total energy of a molecule be shared between its various modes of motion? Maxwell insisted on the principle of the "equipartition of energy"; that is, that the total kinetic energy of a particle should be shared *equally* between the various independent ways (known as "degrees of freedom") in which it could move. A point mass, for example, could evidently move independently along each of three perpendicular axes x, y, and z (or up-down, backwards-forwards, and left-right): It had three degrees of freedom, and one would expect its kinetic energy to be shared equally (on average) between those three directions. But a diatomic molecule could also *rotate* about three independent axes; did not the equipartition of energy require that the same amounts of kinetic energy should be absorbed by the three modes of rotation as by the three modes of translation? On this basis the specific heat of a diatomic gas would be double the specific heat of a monatomic gas, and γ should be 4/3 or about 1.33. By this date it was widely supposed that oxygen and many other common permanent gases were diatomic, but precise and reliable measurements of γ for these gases gave values of about 1.40.

Boltzmann suggested that there must be no significant rotation about the axis joining the two atoms, so that there were in fact only *two* additional degrees of freedom in a diatomic molecule; this indeed gave the correct value of 7/5 or 1.40, but appeared to be an arbitrary constraint. Boltzmann was correct, in fact, but the reason for the validity of the proposed restriction only became clear with the development of quantum theory early in the twentieth century; in yet another important paper in 1907 Einstein showed that there would not be enough energy at room temperature for the higher-frequency axial rotations. Meanwhile the issue of specific heats and the equipartition of energy remained something of a blot on the kinetic theory.

The Less than Ideal Behavior of Real Gases

The early development of the kinetic theory was largely based on the simplifying assumption that gas particles were widely separated and only interacted relatively rarely in occasional momentary collisions; as we have seen, this model was quite successful in predicting many of the properties of "permanent" gases. From the 1870s onwards attempts were made to develop more sophisticated models that might account for aberrations from ideal-gas behavior.

The original simplifying assumption was equivalent to "Mayer's hypothesis," the proposition that, so long as it did no external work, the internal energy (and thus the temperature) of a gas would be independent of its volume. This hypothesis, as we have seen above, was experimentally confirmed—crudely, at least—by Joule's twin-cylinder experiment in 1844. However, more sensitive "porous plug" experiments conducted by Joule and Thomson in 1852 revealed that for most gases there *was* in fact a slight drop in temperature due simply to expansion, indicating that the internal energy of the gas was not entirely independent of its volume. An obvious explanation was that there was a significant degree of attraction between the particles of a gas, requiring some energy to overcome as the gas expanded.

During the same period, there was considerable detailed experimental investigation of the precise behavior of real gases, which established that even such permanent gases as oxygen, nitrogen, and hydrogen deviated slightly from Boyle's law, especially at high pressures (or densities). Regnault himself had conducted experiments since 1847 at pressures of up to 14 atmospheres. Between 1869 and 1893 the French physicist Emile Amagat (1841–1915) performed measurements at pressures eventually reaching 3,000 atmospheres; an early apparatus involved a 300 m column of mercury installed in a disused mineshaft, thereby reaching pressures of 400 atmospheres. In some ways more revealing were studies of the behavior of nonpermanent gases at more normal pressures. From 1863 the Belfast chemist Thomas Andrews (1813–85) performed a series of fundamental investigations of the properties of carbon dioxide (CO_2). Andrews' experiments revealed much about the process of change of state from gas to liquid (and vice versa), establishing the existence of a characteristic "critical temperature" (31.1°C for CO_2) above which a given gas could not be liquefied by pressure alone, thus explaining why some gases seemed to be "permanent" and had resisted every attempt to liquefy them. Andrews suggested that below the critical temperature a gas should be called a "vapor."

Against this experimental background, in 1873 the young Dutch physicist Johannes Diderik van der Waals (1837–1923), in his doctoral dissertation, "On the continuity of the gas and fluid states," proposed a more complex version of the ideal-gas equation that would accommodate these deviations. Van der Waals' new equation of state, instead of the basic $PV = RT$, was

$$(P + a/V^2) \cdot (V - b) = RT$$

The term "a/V^2" was designed to allow for the effect of intermolecular attractions, and the term "b" for the finite sizes of the molecules. The terms "a" and "b" were constants to be determined experimentally. (If $a = b = 0$, then the equation reduced to the familiar $PV = RT$.) The equation not only charted the relatively small deviations of the permanent gases, but the resultant cubic curves predicted very neatly many of the features of Andrews' experimental measurements on carbon dioxide.

In Fig. 7, p. 22, where the values of *pv* observed by Amagat are plotted against the pressures, the process of liquefaction is also illustrated, the vertical portions of the isothermals at and below 30° indicating constancy of pressure with change of *pv* between definite limits.

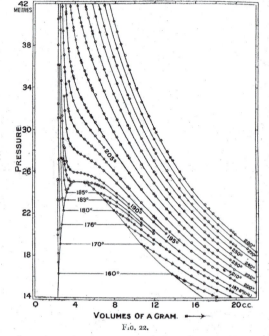

Fig. 22.

In the "Andrews" diagram it will be noticed that the 13·1° isothermal between B and C is not perfectly horizontal, but that it slopes very gently upwards and is rounded off at C, and these peculiarities are even more noticeable at 21·5°. The corresponding portions of the isothermals (0° to 30°) in Amagat's diagram are made perfectly vertical, and in Fig. 22, in which the volumes of a gram of isopentane are mapped against the pressures, the experimental

Figure 6.3: A composite graph of the non-ideal behavior of the hydrocarbon isopentane at different temperatures, with critical temperature 187.8°C. Sydney Young, *Stoichiometry* (London, 1918); by permission of the Syndics of Cambridge University Library.

The Early Reception of the Kinetic Theory

Overall the successes of the kinetic theory were generally thought to outweigh its limitations and inconsistencies, and it was increasingly widely accepted in the 1860s and 1870s. As early as 1861 at a meeting of the American Academy of Arts and Sciences the theory was enthusiastically supported by the American astronomer Simon Newcomb (1835–1909); Newcomb went on to become one of the most eminent American scientists of his day. "One of the most beautiful hypotheses ever promoted in physics," he claimed, "is that which has lately been known as the Dynamical Theory of Gases" (Brush, 1976, p. 203), and, as recorded at the start of this section, he was confident that "the small discrepancy between the observed and computed ratios of the specific heats" would soon be remedied. In 1872, Rayleigh stated that the kinetic theory "has certainly great and increasing claims to be considered at least a truthful representation of the facts" (Rayleigh, 1872, p. 219; Brush, 1976, p. 201). According to Brush (ibid.), Maxwell's lecture on "Molecules" to the 1873 meeting of the BAAS gave "the overall impression...that the kinetic theory was now completely established, and was the most powerful tool available for probing the invisible world of atoms and molecules."

After Maxwell's untimely death in 1879 there was some slackening of productivity and enthusiasm in the field. In the last decades of the century, there was in fact a marked reaction against ruthlessly mechanical, materialistic theorizing in physical science, and there was some revival of the kind of early nineteenth-century skepticism about the reality of atoms that had initially greeted the theories of Dalton. One particular question fueled and focused this skepticism, the question of the exact relationship between the kinetic theory and thermodynamics, and especially the question whether it was possible to provide a rigorous, mechanical explanation for the Second Law.

THE SECOND LAW OF THERMODYNAMICS AND STATISTICAL MECHANICS

The most important question, perhaps, of contemporary scientific philosophy is that of the compatibility or incompatibility of thermodynamics and mechanism. (Bernard Brunhes [1867–1910], at first International Physics Congress, Paris, 1900; translated in Brush, 1976, p. 543)

By the mid-1860s reasonable agreement had been reached on the structure and application of the new science of thermodynamics, and there was widespread acceptance of the kinetic theory of gases. But it was not altogether clear how these two theories should fit together. Whereas the First Law of Thermodynamics and the principle of the conservation of energy were regarded as largely unproblematic generalizations of traditional mechanics, there was much variation in the formulation and interpretation of the more esoteric Second Law (see the chapter 5 sidebar of various versions). It was not clear whether the Second Law was an independent, empirical law of nature, or whether it could be explained in terms of the underlying structure of matter and Newtonian mechanics. The statistical and mechanical methods of the kinetic theory seemed to offer a promising approach towards a reduction of thermodynamics to purely material, mechanical terms. But problems remained and, while some German theoretical physicists strove towards a complete mechanical reduction, the North British group generally retained a degree of privileged independence for the Second Law. Only in the early years of the twentieth century was some degree of consensus reached that reduced the bulk of thermodynamics to the new science of "statistical mechanics," although even then some loose ends and paradoxes remained unresolved.

Maxwell's Demons: British Opposition to German Mechanism

The North British reluctance to see thermodynamics explained in mechanical terms had several causes. On the one hand, there was the traditional British suspicion of speculative hypotheses, a preference for solid, empirical results.

On the other hand, the North British group had deep-seated religious reasons for resisting mechanical reductionism. In general terms, they wished to limit the scope and success of the assertively materialist "scientific naturalism" of Huxley and John Tyndall (1820–93), with its agnostic and atheist tendency. In particular, they wanted to preserve the directional—and thus purposive, even providential—structure of nature that the Second Law seemed to embody.

Within the North British group, two major arguments were developed to clarify the relationship between the Second Law of Thermodynamics and the kinetic theory. Both arguments were initially proposed by Maxwell, who had been contemplating the fundamental status and significance of the Second Law since the late 1850s, and in their original formulations illustrate well his delightfully playful imagination. One argument, which will be discussed below, concerned the apparent contradiction between the time-symmetry of mechanics and the "progressive" directionality of the Second Law. A more original, or even eccentric, argument involved the "thought-experiment" or paradox (or parable, maybe) of what soon came to be known as "Maxwell's demon."

This paradox was first published in Maxwell's 1871 *Theory of Heat*, but had been developed in earlier correspondence with Tait and was originally intended to show that the Second Law had "only a statistical certainty." The experiment involved two equal insulated containers, A and B, supposed to contain equal numbers of molecules at the same temperature, which is to say that the molecules in each container had the same average kinetic energies; the two containers were separated by an insulating partition or diaphragm. "Now," continued Maxwell,

> conceive a finite being who knows the paths and velocities of all the molecules by simple inspection but who can do no work except to open and close a hole in the diaphragm by means of a slide without mass.
>
> Let him first observe the molecules in A and when he sees one coming the square of whose velocity is less than the mean sq[uare] vel[ocity] of the molecules in B let him open the hole & let it go into B. Next let him watch for a molecule in B the square of whose velocity is greater than the mean sq. vel. in A and when it comes to whole [sic] let him draw the slide & let it go into A keeping the slide shut for all other molecules.
>
> Then the number of molecules in A & B are the same as at first but the energy in A is increased and that in B diminished, that is the hot system has got hotter and the cold colder & yet no work has been done, only the intelligence of a very observant and neat fingered being has been employed. (Letter to Tait, 1867; Maxwell 1990–2002, 2, pp. 331–2; Smith, 1998, p. 251)

In other words, by dint of the action of this demon (or, as Maxwell preferred, "valve like that of a hydraulic ram"), *without any work being done*, heat would flow from one container to the other, creating a temperature difference where none had previously existed. Just like a kettle spontaneously boiling on a cold hob, this was precisely what was supposed to be impossible according to the Second Law of Thermodynamics.

Figure 6.4: Maxwell's demon paradox: "a very observant and neat fingered being" could separate fast and slow molecules into containers A and B respectively, thus creating a temperature difference, apparently without having had to do any work, and thereby contradicting the Second Law of Thermodynamics. Illustration by Jeff Dixon.

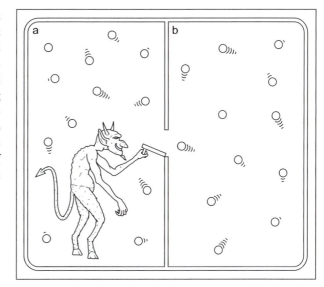

What had been Maxwell's motives in summoning his demon? Was it his purpose to question the universal validity of the Second Law, even to contrive a mechanism that would exploit a loophole and create a perpetual motion machine "of the second kind" to extract heat from the environment without doing work? Not at all. Maxwell explained his motives in a typically light-hearted parody of a Presbyterian catechism—which also contained a dig at Thomson, who had been responsible for re-labeling Maxwell's "finite being" as "Maxwell's Demon":

Concerning Demons.

1. Who gave them this name? Thomson.
2. What were they by Nature? Very small BUT lively beings incapable of doing work but able to open and shut valves which move without friction or inertia.
3. What was their chief end? To show that the 2nd Law of Thermodynamics has only a statistical certainty (Maxwell, in Knott, 1911, pp. 214–5; Smith, 1998, p. 252).

In other words, the Second Law was valid for all practical purposes; strictly speaking, however, this was only true when dealing with "real bodies of sensible magnitude" that are composed of very large numbers of individual molecules. Conversely, "it is continually being violated, and that to a considerable extent, in any sufficiently small group of molecules belonging to a real body" (Smith, 1998, p. 266).

In 1874 Thomson went so far as to calculate the exact probability that a sealed glass jar containing a mixture of two trillion oxygen molecules and eight trillion nitrogen molecules would spontaneously separate into two pure components. "What is the number of chances against one that all the molecules of oxygen and none of nitrogen shall be found in one stated part of the vessel equal in volume to 1/5th of the whole? The number expressing the answer in the Arabic

The Fortunes of Maxwell's Demon

It is unlikely that Maxwell ever regarded his demon as creating a practical loophole in the Second Law. Despite his suggestion that one should speak not so much of a "demon" but rather of "a valve like that of the hydraulic ram," he never suggested that it might constitute the foundation of a perpetual motion machine "of the second kind" with which to extract limitless heat from the cold environment. For many years Maxwell's demon remained merely an intriguing and unresolved curiosity. Studies in the mid-twentieth century, however, by the Hungarian-American nuclear physicist Leo Szilard (1898–1964) and by the French physicist Léon Brillouin (1889–1969) reworked the thought-experiment taking into account the *information* required by the demon to carry out her duties. As explained by Pippard:

> On either side of the adiabatic wall [dividing the containers A and B] the temperature is uniform, and in consequence the radiation in the vessel is isotropic. Therefore the demon cannot distinguish the form or position of any object in the vessel, and cannot tell when to open or close the door. [S]he must be provided with a small flash-lamp to illuminate the oncoming molecules, and this flash-lamp, since it must give out radiation different in character from that in the vessel, necessarily operates irreversibly. Brillouin shows that however well designed the flash-lamp may be, the entropy it generates always exceeds the decrease due to any segregation of molecules achieved with its aid....(Pippard, 1957, pp. 99–100)

This combination of entropy and information and cybernetics—Brillouin coined the term "negentropy" for information—was widely supposed to have resolved the demon paradox—in favor of the Second Law. More recent work, connecting the problem to quantum theory and especially the possibilities of quantum computing, has in turn cast some doubt on this conclusion. Leff and Rex, 1990, provide extensive documentation and discussion of the demon's pilgrimage.

notation has about 2,173,220,000,000 of places of whole numbers" (Thomson, 1874; Brush, 1965, ii, p. 183). (It would thus need some two million or more books of this size in which to write down the odds.) So, while there is a theoretical possibility that the oxygen and nitrogen will spontaneously separate to opposite ends of their container, the likelihood is cosmically small. This view would be accepted in modern thermodynamic theory.

Thomson and Maxwell also developed another apparent paradox intended to clarify the relationship between the Second Law and Newtonian mechanics. Basic Newtonian dynamics was entirely symmetrical with respect to time; any "conservative" event—especially the elastic collision of two atoms—would be *reversible,* would obey the laws of dynamics whether played forwards or backwards in time. The Second Law, on the other hand, the "universal tendency to dissipation," was *directional:* Real irreversible processes were definitely not the same played backwards. How could it be possible to deduce the directional Second Law from reversible dynamics, as Boltzmann claimed to do with the H-function?

As early as 1857 Maxwell had posed the problem in less-abstract terms in a letter to Thomson. Imagine some irreversible event, such as the pouring of water from some jug or vessel into a basin; the motion of the water

> will break up into eddies innumerable....If after a given time say 1 hour you reverse every motion of every particle, the eddies will all unwind themselves till at the end of another hour there is a great commotion in the basin, and the

water flies up in a fountain to the vessel above. But all this depends on the *exact* reversal for the motions are *unstable* and an approximate reversal would only produce *a new set of eddies multiplying by division.* (Maxwell, 1990–2002, 1, pp. 560–1; Smith, 1998, p. 248)

This theoretical reversibility of dynamics allowed Maxwell later to imagine an "Alice Through the Looking-Glass" scenario wherein everything happens backwards:

Now one thing in which the materialist (fortified with dynamical knowledge) believes is that if every motion great & small were accurately reversed, and the world left to itself again, everything would happen backwards: the fresh water would collect out of the sea and run up the rivers and finally fly up to the clouds in drops which would extract heat from the air and evaporate and afterwards in condensing would shoot out rays of light to the sun and so on. Of course all living things would regrede [regress] from the grave to the cradle and we should have a memory of the future but not of the past.

The reason why we do not expect anything of this kind to take place at any time is our experience of irreversible processes, all of one kind, and this leads to the doctrine of a beginning & an end instead of cyclical progression for ever. (Letter, 1868 [1990–2002], 2, pp.360–1; Smith, 1998, p. 239).

Such a bizarre course of events, completely contrary to all human experience, would nevertheless in no way contradict the axioms of Newtonian mechanics, nor the First Law of Thermodynamics.

From a modern perspective the demon paradox and Thomson's calculation of the odds of mixtures of atoms spontaneously separating seem to be designed precisely to provide a basic model for the mechanical *explanation* of the Second Law. In fact, however, it seems that Maxwell and Thomson's purpose was rather to establish the fundamental *independence* of the Second Law. For them, the demon and reversibility paradoxes demonstrated rather that the Second Law could *not* be reduced to Newtonian mechanics. In mechanical terms the universe could quite happily run backwards, in flagrant contradiction of the Second Law, without ever contravening the First Law's conservation of energy. The efforts of Clausius, Boltzmann, and other German physicists to construct a rigorous mathematical explanation of the Second Law in terms of the First must logically be in vain. For the North British group, therefore, the Second Law retained its independent validity; in contradiction of "the materialist (fortified with dynamical knowledge)," it served as the guarantor of direction and "dissipation" in the world, underpinning "the doctrine of a beginning & an end instead of cyclical progression for ever."

In fact, the demon paradox and the "only statistical" validity of the Second Law seems to have meant for Maxwell that the Second Law was ultimately just a reflection of human aspirations and limitations. As he explained:

Available energy is energy which we can direct into any desired channel. Dissipated energy is energy which we cannot lay hold of and direct at pleasure, such as the energy of the confused agitation of molecules which we call heat. . . . [The] notion of dissipated energy could not occur to a being who could not turn any of the energies of nature to his own account, or to one who could trace the motion of every molecule and seize it at the right moment. It is only a being in the intermediate stage, who can lay hold of some forms of energy while others elude his grasp, that the energy appears to be passing inevitably from the available to the dissipated state. (Maxwell, 1890, 2, p. 646; Smith, 1998, p. 240).

To put it in slightly different terms: "The second law relates to that kind of communication of energy which we call the transfer of heat as distinguished from another kind of communication of energy we call work." But if one imagined "our senses sharpened to such a degree that we could trace the motions of molecules as easily as we now trace those of large bodies . . . the distinction between work and heat would vanish" (Maxwell, 1878, p. 279; Smith, 1998, pp. 265–6). With such heightened perceptions, it seemed, the constraints of the Second Law would vanish too. To animals and to angels equally, therefore, the "universal tendency to dissipation," and thus the Second Law, would be irrelevant; it would only be relevant to beings—such as humans—"in the intermediate stage." The Second Law was somehow a result of the demands and limitations of human will and intelligence, the very feature that had originally given "universal dissipation" a Calvinist resonance for Thomson.

Boltzmann: Entropy and Probability

German theoretical physicists seemed not to share Maxwell's reservations. Clausius had been eager from the very start to develop mechanical and molecular explanations for thermodynamic relations; in particular, he sought to explain the entropy of a system in terms of a combination of its temperature and its "disgregation" (that is, the configuration of its constituent particles). The same general approach was vigorously pursued by Boltzmann, using the new statistical methods pioneered by Maxwell.

As already noted, in his 1872 paper Boltzmann had claimed to derive from strictly mechanical foundations a function, later known as the H-function, which showed that, starting from anything other than the Maxwell equilibrium distribution, the net effect of collisions between gas particles would be an increase in entropy; this theorem was presented as a demonstration of the universal, mechanical necessity of the Second Law. A couple of years later Maxwell's reversibility arguments were brought to Boltzmann's attention by his Viennese colleague Josef Loschmidt (1821–95), and they stimulated Boltzmann to further investigation of the Second Law, recorded in a series of papers from 1877 onwards.

The Second Law, Boltzmann argued, could be presented as a tendency of any system to progress from a nonuniform state to a uniform state. "Let us imagine an exceedingly large but not infinite number of perfectly elastic spheres that move in a totally closed container, the walls of which are completely immovable and likewise perfectly elastic. No external forces are supposed to act upon the spheres. At time zero let the distribution of the spheres in the container be non-uniform; for example, let the spheres be more dense on the right than on the left, let them be faster at the top and slower at the bottom, and so on" (Boltzmann, 1909, 2, p. 119). The Second Law required that such a nonuniform distribution would more-or-less rapidly degenerate into a distribution with uniform density and temperature (average kinetic energies), and, once achieved, a uniform distribution would not spontaneously organize itself into a nonuniform state.

The underlying reason for this behavior, Boltzmann reasoned, was quite simply that, if the components of a system (the spheres, in his example above) were allowed to distribute themselves at random, then there would be *many, many more* arrangements that would be more-or-less uniformly disordered than there would be ordered arrangements. In a game of cards there are always many more disordered, losing hands than there are ordered, winning hands. In poker, for example, a desirable, high-value hand (a flush: 9, 10, J, Q, K♥, for example) is high value because it is rare. But it is actually no rarer than any *particular* low-value hand (such as 6♥, 10♠, 9♣, 7♦, K♥); it is just that there are many, many more such low-value hands than there are flushes. The total number of possible five-card hands is, in fact, 2,598,960, but there are relatively very few high-value hands. The high-value hand is rare because it is narrowly defined and highly organized. The normal course of thermodynamic events is represented by supposing that you start with a good hand and randomly replace the cards in it with other cards drawn from the pack at random; the good hand would quite rapidly deteriorate into a series of increasingly poor hands—just because there are so many more poor hands than good ones. (In this scenario, Maxwell's demon would correspond to a card-sharp who could choose both the cards to discard and the cards to select from the pack.) Starting from an ordered, nonuniform state, therefore, and given some process of constant random rearrangement, it was not mechanics but mere *probability* that caused the system to spend most of its subsequent time in one of the much more common uniformly disordered states.

Thus, Boltzmann realized, one could *not* give a mathematical, mechanical *proof* that a system would inevitably progress from ordered to disordered, nonuniform to uniform as the Second Law required.

The following must therefore be considered: a proof that after the elapse of a certain time t_1 the mixture of the spheres must with absolute necessity be uniform, whatever the distribution of state at the beginning of the time may have been, cannot be provided. This is evident simply from probability calculus alone; for each distribution of state (no matter how non-uniform) is, although in the highest degree improbable, yet not absolutely impossible. Indeed it is clear that each particular uniform distribution, resulting from a particular initial state after the lapse of a particular time, is every bit as improbable as any particular non-uniform distribution—just as in the game of Lotto each particular five-card hand is just as unlikely as the hand 1, 2, 3, 4, 5. The greater probability that, as time passes, the distribution of state becomes uniform stems only from the fact that there are many more uniform than non-uniform state distributions. (Boltzmann, 1909, 2, p. 120)

By the same token, one could not *entirely* exclude the possibility that a uniform, high-entropy state would not spontaneously at some point organize itself into a nonuniform, low-entropy state. If one carried on for long enough playing poker in the random, unselective way described above, one would eventually recover a good hand again; similarly, in thermodynamics, according to Boltzmann, "it can be considered impossible that oxygen and nitrogen might initially be mixed in a container in such a way that after the passage of a month chemically pure oxygen separates in the bottom half of the container and nitrogen in the upper half, although according to probability theory this is only exceedingly improbable but not actually impossible." (Boltzmann, 1909, 2, p. 120).

The solution to the reversibility paradox, therefore, lay in recognizing the critical importance of the specific initial condition of "exactly reversed velocities." It was not physically and mechanically impossible that any given system would surrealistically organize itself—that a kettle would spontaneously boil, etc. As Boltzmann realized, the direction in which a system evolved—from nonuniform to uniform or vice versa—depended not just on the laws of mechanics or the H-theorem but critically on the precise initial conditions. The "exactly-reversed-motions" setup in which the entropy would decrease was in fact possible, but it was a rarity amongst billions and billions of alternative scenarios in which the entropy would increase or remain constant. It was the relative frequency of increasing and decreasing scenarios that was significant.

As we have seen above, Maxwell certainly recognized that spontaneous local fluctuations could result in fleeting, local decreases in entropy, and Thomson in 1874 had already calculated the chance of a container of air spontaneously separating into sections of pure oxygen and pure nitrogen. But Boltzmann was more systematic and realized that one needed to consider the complete range of possible arrangements. In the first 1877 paper he merely remarked that "One could even, from the relative numbers of the different distributions of states, calculate their probabilities, which would perhaps lead to an interesting method of calculating thermal equilibrium." (Boltzmann, 1909, 2, p. 121). Within the year this had developed into the famous and fundamental relationship between entropy and probability embodied in the formula $S = k.\log W$, where S is the entropy of a system measurable on the large scale (an aspect, like pressure or temperature, of its "macroscopic" state or, in modern terms, "macrostate"), and W—subsequently known as the "thermodynamic probability"—is a measure of the relative number of possible distinct molecular configurations (or "microstates") that could correspond to the large-scale state. The constant, k, is now known as Boltzmann's constant. A state with low entropy, S, was thus a state with a small relative probability, W, at the molecular level, in the sense that there were relatively few molecular configurations (given a fixed amount of molecules and energy) that could embody that state. This approach to entropy had several advantages. On the one hand it provided a more concrete, physical model to help understand the previously empirical and rather obscure thermodynamic concept of entropy. On the other hand it allowed a definition of the entropy of nonequilibrium systems, something impossible in traditional thermodynamics. Eventually Boltzmann's formula came to be used as the basic definition of entropy, but only after considerable further debate.

Thus it seems that for Boltzmann too the Second Law had "only a statistical certainty," but a certainty that could be explained in terms of reversible mechanics. For Boltzmann there was no fundamental gap between the measurable, human scale and the hypothetical, invisible, molecular scale. Indeed, there should be no absolutely fundamental tendency to the maximization of entropy; rather, the entropy of a system would fluctuate randomly close to its maximum. Fluctuations in the entropy of an isolated system would in fact be symmetrical in time—on average, increase in entropy would be as likely looking backwards as forwards in time. The apparently universal prevalence of dissipation (increase in entropy) was simply a consequence of the fact that we inhabit a particularly organized (low-entropy) bubble within a wider cosmos of disorganized uniformity—we seemed to have been dealt a very good hand, cosmologically speaking. This was hard to square with actual human experience and shifted the debate onto the cosmological scale in the late nineteenth century.

The Heat Death of the Universe, Eternal Recurrence, and Time's Arrow

The Moving Finger writes; and, having writ,
Moves on: nor all thy Piety nor Wit
　　Shall lure it back to cancel half a Line,
Nor all thy Tears wash out a Word of it.
Edward Fitzgerald, The Rubaiyat of Omar Khayyam

The Second Law of Thermodynamics, interpreted as Thomson's "universal tendency to dissipation," provided a direction and inevitable end (and maybe a purpose) to creation. Helmholtz and Clausius endorsed the implication that the universe was ultimately condemned to "eternal, unchanging death." But the spirit of the third quarter of the nineteenth century, especially in Britain, was generally positive and progressive, thanks to industrial and imperial expansion, a spirit reflected in Darwin's theory of evolutionary progress. The common tendency among natural philosophers was to see thermodynamic dissipation as probably part of some wider cyclic process of regeneration.

The last quarter of the nineteenth century, however, was marked by a widespread preoccupation with degeneracy and decadence—the growth of "eugenic" anxieties about national and racial health especially reflect this. In part this may have been a reflection of the "Long Depression" of the European economy from the mid-70s to the mid-90s. This more negative mentality inclined rather to embrace the ultimate "heat death" of the universe. H. G. Wells (1866–1946) is recognized as one of the founders of the then-new genre of "science fiction"; in his *The Time Machine* (1895) the hero is able to fast-forward to the end of the world:

> So I travelled, stopping ever and again, in great strides of a thousand years or more, drawn on by the mystery of the earth's fate, watching with a strange fascination the sun grow larger and duller in the westward sky, and the life of the old earth ebb away. At last, more than thirty million years hence, the huge red-hot dome of the sun had come to obscure nearly a tenth part of the darkling heavens....But I saw nothing moving, in earth or sky or sea....The darkness grew apace; a cold wind began to blow in freshening gusts from the east....From the edge of the sea came a ripple and whisper. Beyond these lifeless sounds the world was silent. (pp. 130–132)

Boltzmann's new probabilistic analysis of entropy, however, seemed to offer a chink through which to escape such dismal prognostications.

Boltzmann's synthesis of entropy and probability required that all possible molecular configurations of a given system were equally probable, just like any particular hand at cards. This seemed to imply that any given system would over time pass through all possible configurations or microstates. Such an assumption became known as the "Ergodic" hypothesis. Partly prompted by a late paper by Maxwell on collections ("ensembles") of similar systems, Boltzmann worked to establish the hypothesis through the 1880s. The idea received some support from work in the 1890s on the so-called "three-body problem" of celestial dynamics by the French mathematician and theoretical physicist Henri Poincaré (1854–1912); Poincaré showed that such a system would eventually return arbitrarily close to any given initial state, although subsequent mathematical analysis has shown that strictly ergodic physical systems are impossible.

NO MORE WATER.

Figure 6.5: The "heat death" of the world, as imagined by the popular science writer Camille Flammarion (1842–1925) in his 1893 novel *La fin du monde* [The end of the world].

Boltzmann's probabilistic interpretation of thermodynamics thus led to another paradox. In a game of poker, even by purely random replacements, one would eventually recover one's original hand of cards. Taken to its logical extreme, the ergodic hypothesis seemed to imply not merely occasional fluctuations in the entropy of a system, but ultimately a return to exactly the original starting point, and not just once, but again and again, albeit over extremely long periods of time. Was the same true for the universe as a whole? Would it also eventually return to its original condition?

The idea of recurrent cycles in the history of the world, involving some degree of periodic return to an original state, is not uncommon in ancient religious and mythological thought. The Babylonian tradition, for example, influencing the Hindus and the Greeks, spoke of a "Great Year" of some 2,160,000 years, after which all the planets would return to their original positions. The notion was revived in the early nineteenth century in poetic and philosophical contexts and was taken up by the German philosopher Friedrich Nietzsche (1844–1900) in the 1880s; Nietzsche was influenced by contemporary popular accounts of recent developments in physics, recasting the old astronomical doctrine in terms of energy and atoms:

> If the world may be thought of as a certain definite quantity of force and as a certain definite number of centers of force . . . it follows that, in the great dice game of existence, it must pass through a calculable number of combinations. In infinite time, every possible combination would at some time or another be realized; more: it would be realized an infinite number of times. And since between every combination and its next recurrence all other possible combinations would have to take place, and each of these combinations conditions the entire sequence of combinations in the same series, a circular movement of absolutely identical series is thus demonstrated: the world as a circular movement that has already repeated itself infinitely often, and plays its game *in infinitum*. (Nietzsche, 1968, p. 549)

For some scientists sympathetic to the antimaterialistic current of the time the recurrence hypothesis served as an argument against the kinetic theory from which it was deduced. Thus, it was argued, eternal recurrence was incompatible with the Second Law of Thermodynamics, and so the kinetic theory must be fundamentally faulty. On the other hand, as Boltzmann was at pains to point out, the time needed for recurrence was so enormously long that the phenomenon could have no empirical significance. Boltzmann perceived, however, that the overall cosmological framework could be of relevance to questions of the nature of the Second Law and of time itself.

Newtonian mechanics, as we have already noted, is symmetrical with respect to time: Not only is there no privileged "present," but there is also no preferred direction forwards or backwards in time, no distinct past or future.

The Second Law of Thermodynamics came to be seen by many as providing precisely that directionality, as a universal tendency either to "dissipation" or to the maximization of entropy. This was questioned in 1897 when, expanding his earlier theorizing about fluctuations in entropy, Boltzmann speculated that our observable cosmos might be only a relatively very small, randomly occurring area of relative order in a vast universe of disorder (or thermal equilibrium):

> There must then be in the universe, which is in thermal equilibrium as a whole and therefore dead, here and there relatively small regions of the size of our galaxy..., which during the relatively short time of eons deviate significantly from thermal equilibrium. Among these worlds the state probability increases as often as it decreases. For the universe as a whole the two directions of time are indistinguishable, just as in space there is no up or down...[A] living being that finds itself in such a world at a certain period of time can define the time direction as going from less probable to more probable states (the former will be the "past" and the latter the "future") and by virtue of this definition he will find that this small region, isolated from the rest of the universe, is "initially" always in an improbable state. (Brush, 1978, p. 70)

In other words, on this basis the Second Law would *define* the direction of time.

But this left many questions unanswered, not least about the precise connection between an increase of entropy and the fundamental human *experience* of the passage of time. Indeed, the issues still remain largely unresolved. Sklar concludes that

> the question of whether an explanatory account of the origin of all of our intuitive temporally asymmetric concepts can be found in the facts about the asymmetry of entropic increase is a real one. It is that question that must be answered in order to judge the correctness of the final Boltzmann thesis. I have argued that although that final thesis is intelligible and far from rejectable as absurd on its face, it is also far from being established...
>
> Anyone who has followed the debate from the days of Maxwell and Boltzmann to the present cannot but help be struck by the way in which the fundamental problems of the theory—the problems posed by its original discoverers and by the brilliant early critics—have remained as deep puzzles for over a century. (Sklar, 1995, pp. 419–20)

Fortunately, the only slightly less profound question of the reality of atoms has been resolved, at least to the satisfaction of most scientists.

BROWNIAN MOTION AND THE
TRIUMPH OF ATOMISM

> I have convinced myself that we have recently come into possession of experimental proof of the discrete or grainy nature of matter, for which the atomic hypothesis had vainly sought for centuries, even millennia....(Ostwald, *Grundriss* 4th ed., 1909; trans. Brush, 1976, p. 699)

Albert Einstein could be said to have placed the capstone on the kinetic theory. In the very early years of the twentieth century the kinetic theory, despite its manifold successes, was on the defensive. No one had ever seen an atom or molecule, and in some scientific circles the kinetic theory was attacked as an unnecessarily complex and speculative explanation of the straightforward experimental observations of thermodynamics. A remarkably successful vindication of the kinetic theory—and of the reality of atoms and molecules—was at hand, however; Einstein's 1905 paper on Brownian motion and the prompt experimental confirmation of its theoretical predictions converted all but the most hard-line skeptics with surprising rapidity.

Brownian Motion and Einstein

The late-nineteenth-century reaction against atomism and the kinetic theory was a complex phenomenon, a combination of cultural, philosophical, and more narrowly scientific factors. It was part of a wider (but by no means universal) reaction against mechanical and materialist physics, most radically exemplified by the "energeticist" school of the German physical chemist Friedrich Wilhelm Ostwald (1853–1932), which regarded *energy* rather than matter as the fundamental reality of the natural world. In 1906 Ostwald affirmed that "as I have been maintaining for the last ten years, the matter-and-motion theory (or scientific materialism) has outgrown itself and must be replaced by another theory, to which the name *Energetics* has been given… Atoms are only hypothetical things…." (Ostwald, 1906, pp. 7, 40–41; Brush, 1976, pp. 698–9). It is significant that Ostwald was a chemist. As we shall see shortly, descriptive thermodynamics had been applied to chemical reactions with great success in the last quarter of the nineteenth century. For some scientists, including the German physicist Max Planck (1858–1947), this marked success contrasted with the increasingly complex and sterile tinkering that seemed to characterize kinetic theory during the same period. The problematic connection between thermodynamics and the kinetic theory, as discussed in the previous section, encouraged further doubts. It was at this juncture that Brownian motion made a decisive contribution.

As recorded at the beginning of this chapter, Robert Brown was one of the most respected British botanists of the early nineteenth century. In 1828 he circulated a pamphlet entitled "A brief account of microscopical observations made…on the particles contained in the pollen of plants; and on the general existence of active molecules in organic and inorganic bodies"; in it he recorded that "While examining the form of these particles immersed in water, I observed many of them very evidently in motion;…These motions were such as to satisfy me, after frequently repeated observation, that they arose neither from currents in the fluid, not from its gradual evaporation, but belonged to the particle itself" (Brush, 1976, p. 658). Such motion or agitation of microscopically small particles had often been observed before, in fact; what was distinctive about Brown's investigation was its extension to include inorganic as well as organic materials, so that the motion was seen to have no particular connection to living organisms. After an initial flurry of controversy, Brown's findings were largely neglected until a slight revival of interest in the 1850s—although none of the major players in the development of thermodynamics mentioned it. In 1879 the German botanist Carl Wilhelm von Nägeli (1817–91) made the first attempt at a quantitative analysis of the phenomenon based on the kinetic theory. He rejected a kinetic explanation, however: On the one hand, he argued, Brownian particles were far too big to be jostled about by molecules, on the other hand the observed speed of their motion was much smaller than the kinetic theory predicted. Debate continued intermittently through the late nineteenth century, and its potential significance for the relationship between thermodynamics and the kinetic theory was clearly highlighted by Poincaré in 1904.

Einstein's analysis of Brownian motion was explicitly directed towards vindicating the atomic-kinetic approach. In his "Autobiographical Notes," written towards the end of his life, Einstein stated that "My major aim in this was to find facts which would guarantee as much as possible the existence of atoms of definite finite size. In the midst of this I discovered that, according to atomistic theory, there would have to be a movement of suspended microscopic particles open to observation, without knowing that observations concerning the Brownian motion were already long familiar" (Schlipp, 1949; Brush, 1976, p. 673). At the time there was no quantitative kinetic theory of *liquids* similar to that developed for gases. Nonetheless, Einstein managed to combine such results as were available and derive an expression for the mean *displacement* of a small particle suspended in a fluid as a function of time. He also derived a formula describing the way that suspended particles should be distributed if allowed to settle under the influence of gravity.

To achieve these results in the absence of a precise kinetic theory of liquids Einstein made a number of bold—and at first sight incompatible—interconnections. On the one hand, he assumed that particles (of whatever size) suspended in a liquid could be treated just like the molecules of a gas; the standard kinetic theory formula could therefore be used to calculate the "pressure" exerted by the particles. On the other hand, there were a couple of precise theories describing aspects of the behavior of fluids. Stokes' formula—describing the force of resistance to the motion of particles of given radius and velocity through liquids of different viscosity—had been established in the 1850s by Thomson's friend George Gabriel Stokes (1819–1903). More recently the Dutch physical chemist Jacobus Henrikus van't Hoff (1852–1911) had developed a quantitative theory of the "osmotic" pressure of a solution in terms of the concentration of the solute. The kinetic theory related largely to gases; Stokes' formula related to macroscopic particles; the van't Hoff theory related to molecules dissolved in liquids. Nevertheless, Einstein's combination produced the required results.

Einstein's original analysis was not altogether easy to follow, but he and others soon presented the findings in more accessible form. He particularly emphasized that his results could readily be checked experimentally; he also pointed out that earlier studies of the *speed* of Brownian motion (as opposed to the displacement) had been misplaced, because the particle would not maintain a velocity in any one direction for more than a split second: "Since we must imagine that the direction and magnitude of these impulses [to movement] are (approximately) independent of the original direction of motion and velocity of the particle, we must conclude that the velocity and direction of motion of the particle will be already very greatly altered in the extraordinarily short time Q (= 3.3 × 10^{-7} seconds), and, indeed, in a totally irregular manner. It is therefore impossible...to ascertain $\sqrt{v^2}$ by observation" (Einstein, 1926, p. 66; Brush, 1976, p. 682). The challenge to provide experimental confirmation was enthusiastically taken up by the French physicist Jean Perrin (1870–1942), who was already engaged in debate about thermodynamics and the kinetic theory. Experiments by Perrin and his students from 1908 onwards did indeed seem to confirm Einstein's predictions (see figure 6.1 above).

The Triumph of Atomism

Perrin was keen to publicize the significance of his results, especially as confirmation that atoms really did exist. In this, although he was not unaided, he seems to have been very successful. As recorded at the beginning of this section, in 1909 Ostwald—much to his credit—completely reversed his previous view on the probable "*discrete or grainy nature of* matter." He went on to explain that "the agreement of Brownian movements with the predictions of the kinetic hypothesis..., which has been shown by a series of researchers, most completely by J. Perrin—this evidence now justifies even the most cautious scientist in speaking of the *experimental* proof of the atomistic nature of space-filling matter" (Ostwald, 1909, preface: trans. Brush, 1976, p. 699).

The rapidity and completeness of the conversion not just of Ostwald but of the bulk of the scientific community is striking. "In fact," Brush (1976, p. 697) remarks, "the willingness of scientists to believe in the 'reality' of atoms after 1908, in contrast to previous insistence on their 'hypothetical' character, is quite amazing. The evidence provided by the Brownian-movement experiments of Perrin and others seems rather flimsy, compared to what was already available from other sources." The answer may lie partly in the strikingly *visual* nature of the phenomenon. In 1909, for example, Hermann Walther Nernst (1864–1941) wrote in relation to Brownian motion: "In view of the *ocular* confirmation of the picture which the kinetic theory provides us of the world of molecules, one must admit that this theory begins to lose its hypothetical character" (Nernst, 1909, p. 212; trans. Brush, 1976, p. 698). So Einstein's analysis resulted in more than just the satisfactory explanation of another empirical phenomenon, such as diffusion or viscosity: Once accepted as a valid explanation, it permitted an almost direct visual perception of the random motions of molecules. In the early twentieth century, therefore, the kinetic theory of gases offered a vivid example of the power of a mechanical, materialist, determinist approach to understanding physical reality.

7

BLACK BODIES, FREE ENERGY, AND ABSOLUTE ZERO

"Nineteenth century clouds over the dynamical theory of heat and light" (William Thomson, Friday evening lecture at the Royal Institution, April 27, 1900, in Thomson, 1904, Appendix B)

By the end of the nineteenth century the phenomena traditionally lumped together under the heading "heat" had been to some extent fragmented among a variety of scientific disciplines—mechanics, electromagnetism, chemistry. Thoroughly to pursue relevant scientific activities in all these areas in the early twentieth century would require another volume as large as the present one. Instead I have chosen to follow a small number of particularly interesting and fruitful strands of development. The reduction of thermodynamics to mechanical principles, culminating in statistical mechanics, has been recorded in chapter 6. In this last chapter we shall study the development of ideas about radiant heat and chemical thermodynamics and we shall finally chronicle the practical struggle to reach temperatures ever closer to absolute zero. The first and last of these processes were intimately involved with the emergence of quantum theory from 1900.

BLACK-BODY RADIATION AND THE QUANTUM THEORY

Planck's treatment of the radiation problem, introducing as it does the conception of an indivisible atom of energy, and consequent discontinuity of motion, has led to the consideration of types of physical processes which were until recently unthought of, and are to many still unthinkable. (Jeans, 1910)

Black-Body Radiation

In 1900, according to William Thomson, Lord Kelvin, there remained two major problems or "clouds" obscuring the dynamical theory of heat and light.

On the one hand, it had proved impossible to detect the "aether," the supposed universal medium for the transmission of electromagnetic radiation. This cloud was dispersed by Einstein's 1905 theory of special relativity. On the other hand, there were problems with the theory of the equipartition of energy within molecules. This second cloud was dispersed by quantum theory, which was initially developed to explain the spectrum of radiant heat and light.

By the early nineteenth century it had come to be recognized that "radiant heat" was significantly different to ordinary heat or, at least, that it constituted a distinct mode of communication of heat, essentially different from the distribution of heat by mixture, conduction, or convection. During the first half of the nineteenth century it was gradually established that radiant heat was in many ways similar in its behavior to light; eventually it was accepted that light and radiant heat (and indeed ultra-violet rays) were just different segments of a broad spectrum of radiations, of which only light was visible. In terms of the wave theory of light, radiant heat came to be labeled "infrared" radiation, having a lower frequency and longer wavelength than visible light. Further investigation of the properties of the radiant heat and light spectrum in the second half of the nineteenth century led to the idea of a universal temperature-related "black-body radiation," and thence, quite unexpectedly, to "quantum" theory, a radical new discontinuous or atomistic understanding of radiation and of energy in general.

Interest in radiant heat in the later nineteenth century was stimulated by a variety of issues, both theoretical and practical. It was a major concern to measure the temperature of very hot bodies, such as the sun and molten metals. Initially, however, it was the development of spectroscopy as a tool of chemical analysis that was probably most influential. Different metals and their salts give off light of characteristically different colors when heated to incandescence. The yellow glow characteristic of sodium, for example, is familiar from street lighting, whereas a pure potassium salt produces a violet color when heated in a flame. This phenomenon resulted in the "flame test" as a simple means of chemical analysis. This technique was given much greater power and precision in the late 1850s by the development of the spectroscope by the German chemists Gustave Robert Kirchhoff (1824–87) and Robert Bunsen (1811–99)—the latter most famous of course for the invention of the ubiquitous "Bunsen burner." The spectroscope used a prism mounted on a turntable to split the light emitted by a heated sample into a spectrum, which could be viewed in detail through a telescope. It emerged that the characteristic colors of the different elements were often produced by narrow lines at specific wavelengths, wavelengths that could be precisely measured on the spectroscope. The yellow light of sodium, for example, turned out to be due not to a band of yellow light but entirely to a pair of very narrow lines very close together in the yellow part of the spectrum. This procedure led to the discovery in the later nineteenth century of several new chemical elements by dint of their distinctive spectra.

Within this context an important issue turned out to be the relative rates at which different bodies absorb or emit radiant heat, an issue that had previously attracted little attention. The French natural philosopher Pierre Prévost

(1751–1839) had suggested in 1792 that thermal equilibrium should be conceived as a *dynamic* balance between heat absorbed from and emanated to a body's surroundings. This "theory of exchanges" was not obviously appropriate within the framework of the fluid caloric theory of heat, however, for the caloric theory saw thermal equilibrium more in terms of a static equalization of fluid levels between different containers. A more experimental approach was taken by the Scottish natural philosopher Leslie, who measured the different rates at which heat was radiated from a metal box—"Leslie's cube"—the sides of which were painted in different colors.

Only in the 1850s, however, did the topic attract sustained attention, largely in response to developments in spectroscopy. It had been observed early in the century that the spectrum of sunlight contained many dark lines, some of which seemed to correspond to the distinctive lines emitted by some elements; for example, the solar spectrum contained a couple of dark lines at the same wavelength as the pair of lines emitted by sodium. Kirchhoff suggested that the dark lines were due to the *absorption* of light from the sun by sodium in the solar atmosphere, and that in general an element would emit and absorb light at the same wavelengths. In 1859 Kirchhoff published a rigorous thermodynamic analysis of the equilibrium radiation within an enclosed container or cavity. It turned out that, whatever the internal surfaces of the container might be, the rates at which they absorbed and emitted thermal (or luminous) radiation had to be strictly linked: At any particular temperature, a good absorber of radiation (at a certain wavelength) had to be a good emitter (at that same wavelength), and a poor absorber must needs be a poor emitter. This became known as Kirchhoff's law. (The Scotch scientist Balfour Stewart had reached a similar conclusion the previous year by a more qualitative argument based on Prévost's theory; the apparently inevitable German-Scottish priority dispute soon followed.) The very best *absorber* of radiation was a black body, one that reflected little or none of the radiation incident upon it. (Hence thermometer

Figure 7.1: A spectroscope, as introduced by Gustav Kirchhoff and Robert Bunsen in the late 1850s. Kirchhoff, *Researches on the Solar Spectrum and the Spectra of Chemical Elements* (Cambridge and London, 1862 [1861]); by permission of the Syndics of Cambridge University Library.

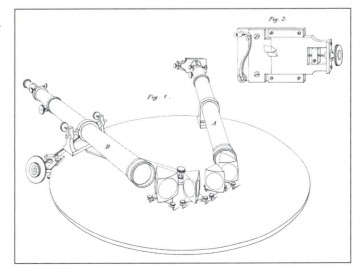

bulbs are sometimes coated with matte black paint in order to maximize their responsiveness to radiant heat.) It followed from Kirchhoff's law, therefore, that the very best *emitter* of radiation would also be just such a black body.

More surprisingly, Kirchhoff was also able to demonstrate theoretically that, once equilibrium had been reached, the intensity of the radiation at different temperatures within his hypothetical enclosed container was a universal function that depended only on the temperature of the container: It did not depend at all upon the nature of the interior surfaces and was identical to the radiation from a perfectly black body at that same temperature. This universal maximum radiation characteristic of a body at a given temperature thus came to be known as "black" or "black-body" radiation. Kirchhoff's theory remained very general, however: It did not specify the precise form of this universal distribution of radiation.

Consequently, there was a concerted experimental and theoretical effort in the last two decades of the nineteenth century both to measure and to explain the exact intensity distribution of black-body radiation and how it varied with temperature. Everyday experience indicated that a hotter body emitted not just more radiation overall, but also more radiation at higher frequencies; the radiation from an iron poker removed "white-hot" from the fire gradually shifts to orange and then red as the poker cools. More precise experimental investigation was encouraged by attempts to measure remotely the temperature of very hot bodies, where traditional thermometers could not be used. The temperature of the sun, for example, had long been of interest to astronomers. The temperature of very hot metals was of equally great interest to industrialists. With the steadily increasing scale of industry, manufacturing processes shifted progressively from tacit craft-based technique to scientific monitoring and control. In 1856, for example, the English engineer Henry Bessemer (1813–98) developed the "Bessemer" converter, the first cheap process for bulk-steel manufacture. To be able to measure the temperature of the molten iron, and thus to optimize its efficient conversion into steel, came to be of huge commercial value.

Serious experimental study of how the intensity of thermal radiation varied with wavelength (at a given temperature) and with temperature was initiated by the American astronomer S. P. Langley (1834–1906). He studied the distribution of thermal radiation from a lamp-black-coated copper radiator that could be maintained at a series of temperatures between 100°C and 815°C. Using a new thermoelectric heat detector or "bolometer" for each temperature he measured the intensity of radiation at different wavelengths. His results were published in 1886. They showed, as might have been expected, that the curves representing the intensity of radiation at different temperatures were similar in shape, but that the total energy emitted (as shown by the areas under the curves) increased with temperature, and that the wavelength of maximum intensity, λ_{max}, at which most heat was given off, steadily shifted towards shorter wavelengths as the temperature increased. His results were broadly confirmed by later researchers, most precisely in 1899 by Otto Lummer (1860–1925) and Ernst Pringsheim (1859–1917) in Berlin.

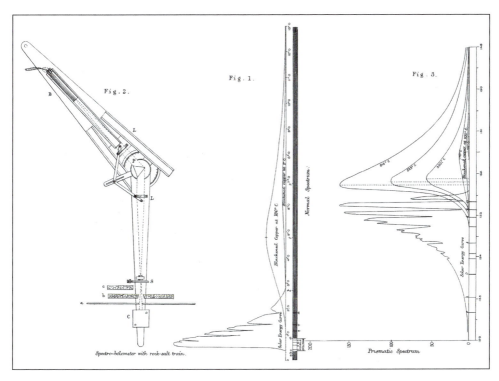

Figure 7.2a: S. P. Langley's 1886 measurements of the intensity of radiant heat at different wavelengths from a series of black-body radiators at temperatures between 10°C and 815°C, compared with the solar energy curve. "Observations on Invisible Heat Spectra..." *Philosophical Magazine*, 21 (1886); by permission of the Syndics of Cambridge University Library.

Figure 7.2b: The distribution of the intensity of black-body radiation at different temperatures according to Lummer and Pringsheim in 1897; the horizontal scale plots the wavelength from 1 to 6 μ (micron, i.e., millionths of a meter), the vertical scale shows the intensity of the radiation. Preston, *Theory of Heat* (London, 1904), fig. 181.

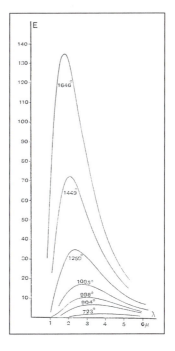

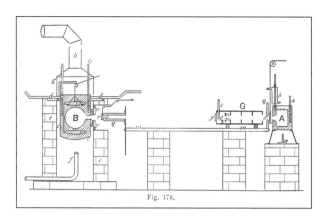

Figure 7.2c: The apparatus devised by Lummer and Pringsheim in 1897 to confirm Langley's measurements; the radiation from the hollow container (B), maintained at temperatures from 100°C to 1300°C, is measured by the sensitive bolometer (G). Preston, *Theory of Heat* (London, 1904), fig.178.

The Ultra-Violet Catastrophe and Quantum Theory

Theoretical efforts to better understand black-body radiation began in the late 1870s, now with the added awareness (thanks to Maxwell) that the radiation of which radiant heat and light were different manifestations was electromagnetic in nature. In 1879 Josef Stefan (1835–93), professor of physics in Vienna, suggested on the basis of some very rough data published by Tyndall in 1864 that the total energy output (E) from a hot body was proportional to the fourth power of its absolute temperature (T), i.e., that $E = \sigma T^4$, where σ (sigma) was a universal constant (now known as Stefan's constant). In 1884 Boltzmann, who was a pupil of Stefan, combined Maxwell's electromagnetic theory with thermodynamics to give a theoretical derivation of the relationship, which is now usually known as the Stefan-Boltzmann law. Within a decade Wilhelm Wien (1864–1928) in Berlin, again using thermodynamic arguments, had derived a connection between temperature (T) and the wavelength of maximum intensity (λ_{max}); according to Wien's displacement law, $\lambda_{max}T$ would be constant, and this was confirmed by the experimental data.

Even so, the precise form of the distribution, as revealed by Langley and others, remained unexplained. Various more-or-less speculative and curve-fitting formulae were proposed through the 1890s. A formula due to Wien appeared to fit the available *high*-frequency measurements well, and was widely accepted. This success was short-lived, however. Early in 1900, new measurements at longer infrared wavelengths revealed the law's inadequacy. An alternative formula was promptly derived by Lord Rayleigh and immediately refined by James Jeans (1877–1946). Based on the assumption of the equipartition of energy between all the different possible frequencies of radiation within a black-body cavity, this formula worked well for the new *low*-frequency measurements. Sadly, however, it predicted an ever-increasing intensity of radiation at higher (ultra-violet) frequencies; Rayleigh's initial purpose had indeed been precisely to discredit the equipartition assumption. This wayward prediction, derived from apparently sound classical principles, came to be known in 1911 as the "ultra-violet catastrophe."

What proved to be the definitive formula was proposed later in 1900 by Max Planck (1858–1947), professor of physics at Berlin. Planck, who was antagonistic to Boltzmann's statistical approach to thermodynamics, had initially been seeking an electromagnetic explanation of the Second Law. This focused his attention on the black-body problem. Albeit based on speculative and somewhat arbitrary foundations—almost indeed as he himself later admitted by "lucky guesswork"—Planck managed to derive a complicated equation:

$$E(\upsilon, T) = 8\pi h \upsilon^3/c^3 . 1/(e^{h\upsilon/kT} - 1)$$

that accurately represented the measured distribution of energy in the black-body spectrum at *both* high and low frequencies. Planck's equation contained a new universal constant, h, now known as "Planck's constant." The equation seemed also to imply that the energy of the radiation at any given frequency (υ)—and in particular the absorption and emission of radiation—was not *continuously* variable, but needed to be treated as though it were restricted to *discontinuous* chunks or "quanta" of precisely defined value ($h\upsilon$). (For those whose Latin is rusty, "quantum" is Latin for "of what size?" and the plural form is "quanta.") Only by virtue of this "chunkiness" could the equipartition assumption that led to the ultra-violet catastrophe be circumvented.

This discontinuity implication, contradicting the classical assumption of *continuous* energy variation, was initially regarded by Planck and most other physicists as a superficial artifact of the mathematics. Gradually, however, during the first decade of the twentieth century, it came to be accepted that this quantum restriction upon the distribution and exchange of energy was fundamental. The background and early development of the quantum theory is described in detail by Kuhn (1978). A crucial role was played by two papers by Einstein who, as we have already seen in the context of Brownian motion, had a far more positive attitude to Boltzmann's statistical methods. Einstein's first paper in 1905 used a quantum approach, treating radiation as a gas of discrete quanta—what would now be called "photons"—to explain the frequency dependence of the photoelectric effect. The second paper, in 1907, treated the absorption and emission of radiation by single atoms and molecules as a quantized process in order to explain the long-standing problems in the theory of the specific heats of gases. The classical assumption of the equipartition of energy between the various modes of motion of polyatomic molecules often gave too high values for their specific heats at ordinary temperatures. Einstein's new account explained this discrepancy and the variation with temperature. New corroborating measurements were at the same time being produced within the burgeoning field of low-temperature physics, and the topic was enthusiastically pursued by the German physical chemist Walther Nernst (1864–1941).

In 1911 the necessity of a radical revision of fundamental thinking about energy in terms of the new "quantum theory" was widely agreed at the international Solvay conference in Brussels (for the organization of which Nernst was largely responsible). Much remained undecided, however: Was it just the processes of absorption and emission that were quantized, or the radiation itself? How could particle-like quanta and the wave nature of radiation be reconciled? The detailed development of the theory would take the best part of another generation.

FROM CALORIC TO CHEMICAL THERMODYNAMICS

I feel very doubtful as to the merits of Willard Gibbs' applications of the "The Second law of Thermodynamics". (W. Thomson, letter to Rayleigh, 1891; quoted in Rayleigh, 1924)

With the decline of the caloric theory and the rise of the dynamical theory of heat and thermodynamics, study of the thermal aspects of chemistry was relatively neglected throughout the middle of the nineteenth century. In the last quarter of the century, however, there was a resurgence of interest, not unconnected with expansion of the chemical industry; this resulted in a vigorous extension of the new thermodynamic principles from the largely thermomechanical domain in which they had originated to a much wider range of phenomena.

From Heats of Reaction to Thermochemistry

As has already been recorded, the precise measurement and proper explanation of the heat of chemical reactions played an important role in the early development of the caloric theory. For Lavoisier, caloric—although "imponderable" or weightless—was to be regarded as a chemical element more or less on a par with other chemical elements, such as oxygen or carbon. Crucially, for Lavoisier, caloric could combine with the other elements and exist in an undetectable "bound" condition, just as oxygen could be undetectably bound up in a metal oxide, for instance. The rival caloric theory due to Irvine and Crawford disputed the existence of "bound" caloric; all caloric was "free" and manifested its presence as a detectable change in temperature. Heats of chemical reaction, therefore, as also heats of melting/freezing and boiling/condensing, were entirely due to differences between the specific heat capacities of reagents and products. The French Institute's 1812 competition to make accurate measurements of the specific heats of gases was partly intended to resolve this dispute. It is not altogether surprising that the outcome was entirely favorable to the French view.

The triumph of Dalton's atomic theory possibly helped to focus attention on the more strictly material (and "ponderable") composition of chemical compounds, and on their structure. Heats of reaction were largely ignored for a generation. There was some revival of interest in the 1840s. In 1840 the Germain Henri Hess (1802–50) proposed what is now known as Hess' law, namely, that the overall heat of formation of a compound from its constituent elements is the same, no matter what route is taken from constituents to compound. A few years later in 1847 this relationship was included by Helmholtz as evidence for his proposed "conservation of force." In the 1850s and 1860s more systematic development of "thermochemistry" took place at the rival hands of the Danish chemist Julius Thomsen (1826–1909) and the very influential French chemist Marcellin Berthelot (1827–1907). Berthelot was responsible for coining the terms "exothermic" and "endothermic" to describe reactions that respectively gave out or absorbed heat. He believed that any spontaneous chemical reaction must necessarily be exothermic. In other words, chemical equilibrium, like the mechanical equilibrium of a system of bodies, was supposed to be determined by the position of minimum energy.

Gibbs and Chemical Thermodynamics

It was not long before attention was drawn to Berthelot's failure to take account of the Second Law of Thermodynamics. According to Rayleigh in an 1875 lecture, "On the Dissipation of Energy," "[t]he chemical bearings of the theory of dissipation are very important, but have not hitherto received much attention" (Rayleigh, 1875, pp. 454–5; Smith, 1998, p. 303). More specifically, Rayleigh was inclined to reject Berthelot's assertion that "the development of heat is the criterion of the possibility of a proposed transformation." How, if that were the case, could one account for endothermic reactions, such as the fall of temperature caused by a freezing mixture?

Unbeknownst to Rayleigh, the comprehensive thermodynamic mathematical apparatus with which to understand chemical processes was already under development by an obscure American professor of mathematical physics, J. Willard Gibbs (1839–1903). From an academic family, Gibbs studied at Yale, producing a doctoral thesis on gears in the engineering school. After three years of further study in Europe, he returned to teach mathematical physics at Yale from 1871; despite being unpaid for the first eight years, he remained there until his death. This initial absence of remuneration did not diminish his productivity. Indeed, Gibbs is probably the most unjustly neglected figure in modern physical science, perhaps because many of his contributions tended to be in the form of general mathematical structures and methods, rather than specific radical theories, such as quantum or relativity theory. His development towards the end of his life of a very general and widely applicable statistical mechanics has already been mentioned, and in a similar vein during the 1880s he developed the enormously versatile tool of vector analysis from the even more abstruse method of quaternions. But his first major achievement was to lay the foundations of a rigorous thermodynamic understanding of physico-chemical equilibria.

During the 1870s he produced a series of powerful papers, culminating in a 300-page memoir, "On the Equilibrium of Heterogeneous Substances." At this point it was still unclear what would determine the equilibrium of complex thermo-mechanical systems. If the entropy were constant, then according to traditional mechanics a system would be in equilibrium in the position of minimum energy—a pendulum will remain at rest at the bottom of its swing. On the other hand, if its energy were kept constant, then a system would reach equilibrium when its entropy was a maximum—one gas diffusing into another until uniformly mixed, for example. Starting from a combination of the first and second laws of thermodynamics in an exact differential form, dU = TdS–PdV, Gibbs showed how to reconcile the competing equilibrium requirements of minimizing energy and maximizing entropy. It turned out that most spontaneous reactions would indeed be exothermic and would achieve equilibrium by giving off energy, just as Berthelot maintained. But this would not always be the case. In a freezing mixture, for example, the increased entropy of the more disordered distribution of the solute outweighs the fact that extra energy is required to affect the solution; the process of solution therefore proceeds even though it is endothermic and energy is *absorbed*, cooling the mixture and its surroundings.

Gibbs went on to apply his basic approach to complex physico-chemical systems involving both different phases (solid, liquid, and gaseous) and mixtures of chemically interacting substances. In the short term, Gibbs' work was largely ignored by European scientists, with the major exception of Maxwell. As already noted, as early as 1873 Maxwell was referring to Gibbs to correct Tait's presentation of entropy; a couple of years later he was vigorously promoting the importance of Gibbs' methods in understanding "the relations between the different physical and chemical states of bodies" (Maxwell,1908 [1876], pp. 819–20; Smith, 1998, p. 262). The style of Gibbs' writing was certainly difficult, even impenetrable, with few attempts to illustrate or explain the basic argument. Even Rayleigh found it hard going, disarmingly advising Gibbs in 1892 that his "Equilibrium" paper had been "too condensed and too difficult for most, I might say all, readers" (Klein, 1972, p. 390; Smith, 1998, p. 303). William Thomson, if his private correspondence with Rayleigh quoted at the beginning of this section is to be believed, had been even less enthusiastic.

It is perhaps not entirely surprising, therefore, that when Helmholtz came to tackle similar issues in 1882 in a memoir on "The Thermodynamics of Chemical Processes" he was apparently unaware of Gibbs' work. He nevertheless arrived at many similar conclusions, perhaps in a more accessible format. Helmholtz showed that the crucial variable was the "free" energy of a system, that is, the amount of energy that was available to perform work. Any system would be in equilibrium when its "free energy," influenced by variations in both energy and entropy, was at a minimum. The precise formula depended upon the constraints upon the system; under conditions of constant temperature and pressure (typical of many chemical reactions), equilibrium is determined by the minimization of what is now known as "Gibbs' free energy," G, defined by the equation $\Delta G = \Delta H - T\Delta S$, which is known, appropriately enough, as the Gibbs-Helmholtz equation.

Helmholtz's enormous reputation ensured that his ideas, and eventually Gibbs', received widespread attention. This was especially the case among the pioneers of the new discipline of "physical chemistry" that was being forged by Ostwald, van't Hoff, and others, who were especially preoccupied with problems of solution, osmotic pressure, and so on. Ostwald would claim for Gibbs that "to physical chemistry, he gave form and content for a hundred years." A more detailed modern assessment of Gibbs' significance is that he "vastly extended the domain covered by thermodynamics, including chemical, elastic, surface, electromagnetic, and electrochemical phenomena in a single system.... [His] memoir showed how the general theory of thermodynamic equilibrium could be applied to phenomena as varied as the dissolving of a crystal in liquid, the temperature dependence of the electromotive force of an electrochemical cell, and the heat absorbed when the surface of discontinuity between two fluids is increased" (Klein, 1972, pp. 389–90).

TOWARDS ABSOLUTE ZERO, OR THE CHANCELLOR'S CHICKEN COMES HOME TO ROOST

If... the earth were taken into very cold regions, to those of Jupiter or Saturn,... [t]he air, or at least some of its constituents, would cease to remain an invisible gas and would turn into the liquid state. A transformation of this kind would thus produce new liquids of which we as yet have no idea. (Lavoisier, quoted in Mendelssohn, 1966, p. 7)

The theoretical existence of an absolute zero of temperature at around –273°C was clearly established by Thomson's new thermodynamic temperature scale. Unfortunately, prior to the mid-nineteenth century there were no effective artificial means of achieving very low temperatures. In the second half of the nineteenth century, however, refrigerated transport became increasingly crucial to the expansion of global agriculture and the sustenance of European urban industrial populations. With this added technical and financial stimulus, engineers and scientists devised increasingly powerful methods of creating ever-lower temperatures. By 1908 scientists had reached temperatures within a few degrees of absolute zero. Along the way they had managed to liquefy all the so-called "permanent" gases. Ironically, the systematic measurement of specific heats at low temperatures confirmed what came to be known as the Third Law of Thermodynamics, namely, that it would be impossible ever to reach absolute zero.

The Liquefaction of the Permanent Gases

In the eighteenth and early nineteenth centuries the only useful means of creating artificially low temperatures was a "freezing mixture," as used by Fahrenheit for example to establish the maximum degree of cold on his thermometers. The cooling effects of the evaporation of volatile liquids (especially ether) and of the expansion of compressed air had both been noted, however. The preservative effect of storing perishable goods at low temperatures had also long been recognized—and had indeed been investigated by the Lord Chancellor Francis Bacon in 1626, with unfortunate consequences. Many eighteenth-century English country houses would have a subterranean ice-house in the grounds, which in winter time would be packed with ice and snow that would last well into the summer. In the early nineteenth century a vigorous trade in natural ice developed, especially in the eastern United States. In the mid-nineteenth century the very rapidly expanding urban populations of Western Europe became increasingly dependent upon a corresponding expansion of agriculture in the Americas, Australia, and New Zealand. However, whereas cargoes of grain would quite readily survive the long journey, meat needed to be chilled or frozen. Since natural ice was inevitably in short supply—especially in Australia—designs for artificial ice-making machines began to be patented from the 1830s onwards. These designs relied either upon the expansive cooling of compressed air or upon the evaporation of a volatile liquid: Ether or ammonia were commonly used. The latter effect is the basis of the modern domestic refrigerator. In a continuous cyclic process a suitable vapor is compressed at room temperature and thereby liquefied; when allowed to expand at low pressure through a throttle valve the liquid evaporates, cooling substantially as it does so; the low-pressure vapor then returns to the compressor and starts the cycle again. The increasing efficacy of refrigerated transport enabled an extraordinary growth in the scale of meat shipments in the last quarter of the nineteenth century. In 1880, for example, some 3,571 carcasses of mutton were successfully shipped from Argentina to London; by 1900 this figure had exploded to over two million, with half a million hundred-weight of beef into the bargain.

Boosted by these technical developments and to no small degree sponsored by the refrigeration industry, low-temperature physics simultaneously blossomed. Throughout the nineteenth century considerable attention was directed to the behavior of gases under pressure, and especially to the liquefaction of gases. Interest was commercial as well as purely academic; oxygen in particular became an essential component of the bulk-steel manufacturing industry and was widely used in the oxy-acetylene torch; a cheap method of extracting oxygen from air was highly desirable therefore. Although many gases could be liquefied simply by compression, several common so-called "permanent" gases—notably oxygen, nitrogen, and hydrogen—stubbornly remained gaseous even under thousands of atmospheres' pressure. The key to this failure was provided by Andrew's experiments on carbon dioxide, from which emerged the concept of a "critical temperature" characteristic of a given gas above which it could not be liquefied, no matter what pressure might be applied. It became apparent that the critical temperatures of oxygen and nitrogen (–118°C and –146°C, as it subsequently turned out) were still well below any degree of cold so far reached.

The way forward was evidently via a combination of high pressures and even more drastic cooling. The liquefaction of oxygen was finally achieved simultaneously and independently in 1877 by the French engineer Louis Paul Cailletet (1832–1913) and the Swiss physicist Raoul Pictet (1846–1929). Both men used the evaporation of less volatile vapors (i.e., those with higher critical temperatures) as a first step towards low temperatures. Using sulfur dioxide as a refrigerant, Cailletet first lowered the temperature of a flask of oxygen to about −29°C. Although the gas had previously been compressed to a pressure of 300 atm., this was still not sufficient to liquefy it. When the pressure was released, however, the expansive cooling of the oxygen reduced its temperature below its boiling point, and a fine mist of liquid droplets was observed, sufficient at least to convince Cailletet (and the French Academy) that oxygen could be liquefied, however fleetingly.

Pictet used a sequence of increasingly volatile gases—sulfur dioxide, carbon dioxide, oxygen—to reach even lower temperatures. Such a sequence or "cascade" of gases was widely used by later scientists. Even so, Pictet's results were as ephemeral as Cailletet's. In 1883, however, a pair of Polish researchers, Szygmunt Florenty Wroblewski (1845–88) and Karol Stanislaw Olszewski (1846–1915), using ethylene (ethene) evaporating under reduced pressure, reached a temperature of some −130°C; this was finally below the critical temperature of oxygen, which could then be liquefied by increased pressure. The Poles were thus able to prepare stable samples of liquid oxygen quietly boiling in a test tube. The slightly more difficult preparation of liquid nitrogen soon followed. Hydrogen, however, defied all attempts to liquefy it. Theoretical calculations based on van der Waals' equation suggested that it had a critical temperature as low as −243°C (30K), significantly lower than could be reached even by the evaporation of liquid nitrogen. Some new technique would be needed to bridge the gap.

The requisite new technique was first developed as a *commercial* process for the manufacture of liquid air and oxygen. By the late 1890s, four-stage cascade apparatus had been developed that could produce substantial quantities

Figure 7.3: The apparatus developed by Carl von Linde in 1895, exploiting the Joule-Thomson effect to produce liquid air in bulk to meet rapidly expanding commercial demand. Edser, *Heat for Advanced Students* (London, 1923), fig. 168.

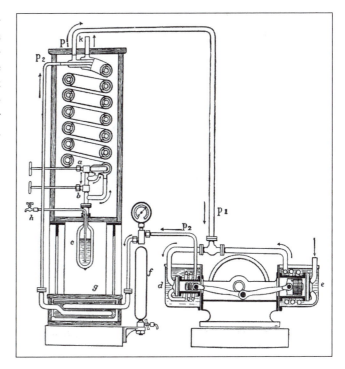

of liquid air for use in the laboratory, but such equipment was inevitably very complex. In 1895, therefore, to meet increasing demand from industry, Carl von Linde (1842–1934) in Germany and William Hampson (1854–1926) in England independently developed a much simpler (in engineering terms) and continuous process exploiting the Joule-Kelvin (or Joule-Thomson) effect. In the Joule-Kelvin effect most gases when they expand, even when they do not perform any *external* work, still cool very slightly because of the *internal* work done against intermolecular attractions. The effect is very slight but, if the expanded and cooled gas is used to pre-cool incoming compressed gas, then the effect can be cumulative or "regenerative." This procedure had the additional significant advantage that it involved no moving parts at very low temperatures, and thus avoided the need for special lubricants.

The race to liquefy hydrogen was won by the flamboyant Scottish scientist (Sir) James Dewar (1842–1932), professor of chemistry at the Royal Institution in London from 1877. He had a very wide range of experimental research interests, ranging from explosives to soap bubbles, but it is for his low-temperature work that he is best remembered. Already in the 1870s he had devised the double-skinned vacuum "Dewar" flask in which to store liquid gases with minimal heat loss—a design that was subsequently marketed as the vacuum or "Thermos" flask in which to store hot drinks. In 1898, using the Joule-Kelvin effect, Dewar succeeded in liquefying hydrogen at

Figure 7.4: Sir James Dewar demonstrating the properties of liquid hydrogen to the scientific and social elite at the Royal Institution in 1904; among those in the audience are the scientists Sir William Crookes, Lord Rayleigh, and Sir Oliver Lodge, the statistician and founder of eugenics Francis Galton, and the radio pioneer Guglielmo Marconi. "A Friday Evening Discourse at the Royal Institution; Sir James Dewar on Liquid Hydrogen," 1904 (oil on canvas), Henry Jamyn Brooks, (1865–1925) / The Royal Institution, London, UK / The Bridgeman Art Library.

about −253°C or 20K. He subsequently went on to solidify hydrogen at about 15K. Somewhat to his surprise, however, he remained unable to liquefy the denser gas helium, which had only recently been isolated in the laboratory. That final stage of the race was won by the Dutch physicist Heike Kamerlingh-Onnes (1853–1926) in 1908, also using the Joule-Kelvin effect to liquefy helium at about 5K.

In stark contrast to Dewar's highly personal style of research, Kamerlingh-Onnes' laboratory may be seen as a prototype of the large, almost industrial "Big Science" research institute that increasingly dominated the physical sciences through the twentieth century. Appointed professor of physics at the University of Leiden in Holland in 1882, Kamerlingh-Onnes devoted his inaugural address to "the importance of quantitative investigations in the physical sciences," crystallized in the motto "Door meten tot weten (Through measurement to knowledge)" (Mendelssohn, 1966, p. 74). His success depended not only upon his understanding of the theoretical and experimental issues, but also upon his ability to coordinate a complex operation: Apart from meticulous long-term planning of each experimental program, Kamerlingh-Onnes set up a school to train the highly skilled instrument-makers and glass-blowers needed to make the complex apparatus; he attracted large numbers of research students to help with the work, establishing a journal exclusively devoted to publishing his department's findings; most significantly, as the scale and cost of apparatus, training, and publishing spiraled, he forged very close links with the refrigeration industry as a source of funding. For the next couple of decades Kamerlingh-Onnes' laboratory dominated low-temperature technology, or "cryogenics" (from the Greek for "frost-making") as it had been christened in the 1890s, and he himself was awarded a Nobel prize in 1919.

Low-Temperature Physics and the Third (and Final) Law

In the next few decades the most important results emerged from the systematic investigation of the properties of materials in the newly accessible low-temperature domain. Nevertheless, the struggle to reach ever-lower temperatures certainly remained a significant goal for its own sake. Having liquefied helium at about 5 K, Kamerlingh-Onnes attempted to reach lower temperatures by evaporating the helium under reduced pressures. Even some 14 years later, however, and using the most powerful vacuum pumps then available, developed for the burgeoning electronics industry, he was unable to achieve temperatures much below 1 K. In the absence of any substance more volatile than helium, the road towards absolute zero appeared to be blocked. Within a few months of Kamerlingh-Onnes' death in 1926, however, an entirely new technique of *magnetic* cooling was suggested. Using the "adiabatic demagnetization" of certain types of magnetic salts, it proved possible in following decades to reach temperatures of a few thousandths of a Kelvin. Using the even more powerful technique of nuclear magnetic cooling, it is now possible to reach temperatures within a few millionths of a degree of absolute zero.

The race to achieve ever lower temperatures, and specifically to liquefy the permanent gases, had dominated late-nineteenth- and early-twentieth-century low-temperature physics. After 1908 Kamerlingh-Onnes directed his energies more towards the study of low-temperature properties for their own sake, and with some astonishing results. Electrical resistance recommended itself as a subject of study, partly because it was relatively easy to measure, even at very low temperatures. Kamerlingh-Onnes expected resistivity to decrease steadily with temperature. However, experiments at Leiden, initially with mercury (because it was easy to purify), revealed a sudden drop in resistivity, apparently to zero, at a sharp cut-off temperature of 4.15 K. Kamerlingh-Onnes had stumbled upon the phenomenon of "superconductivity." Initially skeptical of his own measurements, he went on to confirm the phenomenon in other metals, including tin and lead. Theoretical understanding in terms of quantum mechanics was only achieved in the 1950s.

When combined with the development of chemical thermodynamics, the growth of low-temperature technology led to a further fundamental addition to the canon of thermodynamic laws. One of Ostwald's students was Walther Nernst. Nernst was struck by the fact that at ordinary temperatures most spontaneous reactions are indeed—as Berthelot maintained—exothermic. This seemed to suggest that the difference between the total energy and free energy of a reaction was usually small at normal temperatures. This difference depended in turn upon the entropy term in the Gibbs-Helmholtz equation. This therefore led Nernst to propose his "heat theorem," namely, that rather than the energy, it was the *entropy* of all pure crystalline solids that tended to zero at the absolute zero of temperature. To confirm Nernst's suggestion required that the total energies and the entropies of substances as calculated according to the heat theorem should be compared with more direct measurements at room temperature. This in turn required the careful measurement of the specific heats of materials from (close to) absolute zero upwards. This gave an added stimulus to a program of low-temperature measurement of specific heats, which confirmed not only Einstein's quantum theory of specific heats but also Nernst's theorem.

The experimental measurements also provided a wealth of data to aid the prediction and manipulation of chemical reactions for practical purposes, such as the manufacture of ammonia from nitrogen and hydrogen. But the heat theorem also had interesting theoretical implications. One consequence of the theorem was that it should be impossible actually to reach absolute zero in a finite number of steps. It is in this vivid but somewhat less useful form that it is often popularly presented. (A detailed but accessible account of the complex theoretical and experimental interactions between Nernst's theorem and quantum mechanics may be found in Mendelssohn, 1966.) Initially the exact status of the theorem remained unclear; the possibility that it might be derivable from the Second Law of Thermodynamics was eventually rejected, and it came to be accepted as an independent "Third Law of Thermodynamics."

Nernst was awarded the Nobel prize for chemistry in 1920. With typical tongue-in-cheek insouciance he argued that his Third Law had to be the final fundamental law of thermodynamics, basing his claim on a thumbnail sketch of the history of the discipline: The First Law, he pointed out, had been discovered by three people (namely, Mayer, Joule, and Helmholtz) and the Second Law by two (namely, Clausius and Thomson); logically therefore the Third Law, discovered by just one person (himself), had to be the last in the series.

CONCLUSION

[B]efore 1850 the steam engine did more for science than science did for the steam engine. (attributed to L. J. Henderson [1878–1942]; see James B. Conant, 1951)

Between the late sixteenth and the early twentieth centuries, then, natural philosophers' and scientists' understanding of "heat" had gone through several phases. Aristotle and the medieval schoolmen saw heat (and cold) in essentially meteorological and physiological terms: Hot and cold and wet and dry were the two pairs of qualities that defined not only the four elements (including hot, dry fire, for example), but also the four bodily humors (including hot, dry "choler," or bile). In the first half of the seventeenth century this qualitative, descriptive, and in many ways common-sense view of natural structures was vigorously challenged both by an enthusiasm for artificial instruments and apparatus (including the thermometer and the vacuum-pump), and by an insistence that all natural phenomena had to be explained in mechanical terms, that is, in terms of the shape, size, and motion of particles of matter. From this perspective, heat was usually explained as a manifestation of the more-or-less agitated motion of the microscopic atoms or corpuscles of which matter was supposed to be made.

By the early twentieth century a metaphysically similar "dynamical" or kinetic view of heat (as a by-product of the motions of atoms and molecules) had finally prevailed—although this view was now based on complex mathematical and statistical theorizing and extensive and precise supporting experiment, as illustrated by the work of Einstein and Perrin on Brownian motion. Despite the best efforts of some British scientists, especially Tait, to draw a neat straight line from Bacon and Locke through Rumford and Davy to Joule and Thomson, it is in fact unrealistic to trace a straightforward historical connection between the speculations of the seventeenth-century mechanical philosophers and the statistical mechanics of the late nineteenth and early twentieth centuries.

A century or more elapsed between the speculations of the seventeenth century mechanical philosophers and the resurgence of the dynamical theory of heat in the 1840s. In the intervening later eighteenth and early nineteenth centuries, for the vast majority of scientists their thinking was dominated by *material* theories of heat. Heat was deemed to be a "subtle," weightless (but nonetheless material) fluid, called "caloric" by Lavoisier and included by him at the head of his list of the chemical elements. This preference for a material view of heat was not a regrettable lapse of scientific judgment. Rather, caloric as an indestructible fluid (as opposed to an ill-defined motion or agitation of invisible particles) allowed heat to be quantified and precisely measured in the newly designed ice-calorimeter. The caloric theory enabled heat to be included within the emerging discipline of mathematical physics.

Enthusiasm for the caloric theory faded only gradually in the second quarter of the nineteenth century. To be successful, a revived dynamical theory of heat would also need secure quantitative foundations, which the earlier mechanical speculations largely lacked. That foundation was provided by the quintessentially nineteenth-century industrial concept of "work," defined—in terms painfully familiar to any navvy or "navigator" who helped to build the canals and railways of the age—as the capacity to lift weight through height. It still needed Joule's series of painstaking experiments to establish that heat could be measured in terms of work, that there was a fixed exchange rate between the two currencies, "a mechanical equivalent of heat." This correlation of heat and work was promptly generalized by William Thomson and his colleagues. "Energy," quantified as the capacity to perform work, was defined as the new universal substratum of all physical phenomena. Quantification, the balancing of the books, was guaranteed by the First Law of Thermodynamics, the conservation of energy, which became the cornerstone of the new discipline of physics, and indeed of modern science.

Looking back over three hundred years of development, it is necessary to consider what factors have driven that process. Science is often presented as an integrated structure of theory and experiment, and scientific progress as the inevitable result of interplay between new experimental results and new theories. Clearly, there is a lot of truth in this; incisive theoretical speculation and precise and painstaking experimentation have each played fundamental roles in the development of ideas about heat. The origin of thermodynamics might be seen precisely as a combination of Carnot's theorizing and Joule's experiments. But this view of the operation of science benefits from the simplification of hindsight. Elegant and powerful theories do not always convince, nor do what come to be seen as "crucial" experiments. Carnot and Joule were both ignored initially. Bernoulli's (correct) derivation of Boyle's law from kinetic principles was largely ignored in favor of Newton's static explanation. Rumford's cannon-boring experiments, subsequently presented as "decisive," were also very largely neglected. And in many ways rightly so: At that point in time the caloric theory offered a more fruitful, quantitative foundation for further research.

Modern science is a collective enterprise. The popular image of the creative scientist as a lone, often misunderstood genius is fundamentally flawed. One of the essential features of the "new philosophy" as it consolidated in the later seventeenth century was its insistence on collective responsibility for knowledge. Robert Boyle and the Royal Society, for example, saw the collective witnessing of the fellows as the best guarantee of experimental honesty and objectivity. This was in tacit contrast to the mentality of the Renaissance magician seeking personal enlightenment (and power) in the privacy and secrecy of his own laboratory. It was in the aftermath of the French Revolution, however, that something like the professional institutional structures of modern science first emerged. French science, albeit dominated by individuals such as Laplace and Berthollet, provided coherent structures for training and careers, funds and facilities for research, journals in which to publish results, and collectively coordinated programs of research. Inevitably, however, such institutions, especially in the hands of a few individuals, can be resistant to innovation. This was evident in the initial reluctance of the Parisian scientific elite to respond to the work of Fourier and Fresnel.

Thus the development of scientific ideas has evidently been a complex matter of theories, experiments, and collective interaction between scientists, procedures largely *internal* to science and the scientific community. But the development of scientific thinking has also been powerfully influenced by a rich complex of *external* influences from the wider culture. These external influences have acted on scientific thinking at several different levels. On a very practical plane, science has often depended upon the development of tools and techniques in the wider culture. The seventeenth- and eighteenth-century development of thermometry, for example, depended on the general growth of glass-working technology. In comparable but more intellectual terms, some of the statistical methods applied to the kinetic theory in the later nineteenth century were drawn from "statistics" proper, that is, the very nineteenth century science of collecting and collating information relevant to the running of a modern bureaucratic state.

At a more general level, the deepest and most pervasive beliefs and models of a culture can also have substantial influence. It has been argued, for example, that the religious commitments of William Thomson and other Scottish scientists contributed to the early development of thermodynamics; their concern with waste and efficiency in general, and a sensitivity to the "universal tendency to dissipation" in nature in particular, both resonated with their Presbyterian beliefs. Similarly, we have seen that an awareness of the fundamental economic significance of capital and currency served to explain and legitimate, if not to create, the new doctrine of "energy" as a universal medium of physical exchange.

But the most powerful external, cultural influences on science have perhaps been more directly industrial and economic. At the turn of the nineteenth century, for example, the development of low-temperature physics was closely connected, both technically and financially, with the burgeoning refrigeration industry. Nowhere are such interactions more evident, however, than in the early construction of thermodynamics. It is very difficult to imagine the emergence of modern thermodynamics without the preceding development of the steam engine. The development of the steam engine itself, although not entirely disconnected from contemporary scientific activity, was primarily elicited by the expanding demands of the Industrial Revolution, especially in Britain. But it is especially from about 1820 onwards, with an ever-increasing emphasis on efficiency, first in France in the aftermath of the Napoleonic wars and then in the context of marine engineering to serve expanding global trade, that the set of problems and preoccupations emerged that lead to thermodynamics. This intimate interaction between industry, commerce, and science was clearly recognized by Sir Joseph Larmor (1857–1942), Lucasian Professor of Mathematics at Cambridge, in his obituary of William Thomson, Lord Kelvin (Larmor, 1908, p. xxix):

> If one had to specify a single department of activity to justify Lord Kelvin's fame, it would probably be his work in connexion with the establishment of the science of Energy, in the widest sense in which it is the most far-reaching construction of the last century in physical science. This doctrine has...furnished a standard of industrial values which has enabled mechanical power in all its ramifications, however recondite its sources may be, to be measured with scientific precision as a commercial asset.

TIMELINE

Date	Event
B.C.	
c. 500	Heraclitus of Ephesus claims that all things are made of fire.
c. 387	Plato founds Academy at Athens.
355	Aristotle founds Lyceum at Athens.
323	Death of Alexander the Great.
c. 300	Euclid's *Elements* lays foundations of geometry.
c. 60	Lucretius develops atomic theory of matter.
44	Assassination of Julius Caesar; Rome becomes an empire.
A.D.	
313	Constantine the Great makes Christianity official religion of the Roman Empire.
410	Sack of Rome by the Goths; terminal decline of Roman Empire in the West.
622	Mohammad's flight to Medina; subsequent spread of Islam in Middle East and around the Mediterranean.
late 12th century	Revival of classical and Arabic learning in Western Europe; emergence of universities.
1318	First record of gunpowder and cannon.
1348	Black Death sweeps Europe.
1453	Constantinople falls to the Turks; end of Roman Empire in the East.
1455	The Bible printed by Gutenberg.

1492	Columbus arrives in the West Indies.
1517	Luther publishes 95 theses on "indulgences" in Wittenberg; start of the Protestant Reformation.
1519–22	Magellan circumnavigates the globe.
1536	Calvin's *Institutes of the Christian Religion*.
1543	Copernicus publishes his sun-centered theory of the solar system.
1607	First permanent English colony founded in North America
1610	Galileo publishes *Starry Messenger*, reporting telescopic observations.
1612	Santorre gives first description of a thermometer.
1620	Bacon's *Magna Instauratio (Great Restoration)* proposes new natural philosophy based on cumulative experiment and observation.
1628	Harvey discovers the circulation of the blood.
1633	Galileo forbidden to teach heliocentric theory by Roman Inquisition.
1642–49	English Civil War; King Charles I beheaded.
1644	Descartes' *Principia Philosophiae* develops mechanical philosophy. Torricelli invents mercury barometer.
1656–66	Accademia del Cimento in Florence invents sealed thermometer.
1660	Restoration of English monarchy. Foundation of Royal Society, refounded with royal charter in 1662. Boyle's *New Experiments Physico-Mechanicall* using the air-pump.
1662	Boyle proposes Boyle's law for gases.
1665	Royal Society begins publication of *Philosophical Transactions*, the first journal of a strictly scientific nature.
1666	French Académie Royale des Sciences founded in Paris by Louis XIV, "the Sun King."
1687	Newton's *Principia Mathematica* develops a new cosmology based on universal gravity; also deduces Boyle's law as a repulsion of static gas particles.
1698	Savery granted patent for steam suction pump.
1699	Amonton investigates thermal expansion of air.
c. 1710	Newcomen develops atmospheric steam pump.
1724	Fahrenheit proposes his temperature scale.
1735	Linnaeus' *Systema Naturae* classifies plant and animals.
1738	Daniel Bernoulli's *Hydrodynamics* deduces Boyle's law from a kinetic theory of gases.
early 1740s	Celsius proposes Celsius temperature scale.

1751–72	Diderot and d'Alembert co-edit the *Grande Encyclopédie*, widely regarded as the most influential work of the eighteenth century.
1759	Smeaton publishes experiments on efficiency of waterwheels.
c. 1760	Black works on specific and latent heat.
1760s	Industrial Revolution in Britain gathers momentum, e.g., opening of Bridgewater canal to take coal to Manchester in 1762; Arkwright's spinning frame mechanizes spinning in 1768.
1769	Watt's patent for a separate condenser.
1772	Wilcke publishes experiments on latent heat.
	Lavoisier experiments on combustion.
1777	Scheele's *On air and Fire* distinguishes radiant heat.
1779	Crawford's *Animal Heat* develops Irvine's theory of latent heat.
1783	First manned flight in Montgolfier brothers' hot-air balloon.
	Lavoisier and Laplace's *Memoir on heat* develops caloric theory of heat and the ice-calorimeter.
1789–94	The French Revolution.
1789	Lavoisier's *Elementary treatise on chemistry* gives systematic account of new oxygen theory of combustion and a new system of chemical nomenclature.
	Lazare Carnot's *Essay on machines in general*.
1793	Reform of French educational and scientific institutions; First Class of the Institute replaces Académie Royale.
1796	Steam indicator invented by Southern.
1798	Rumford publishes experiments on generation of heat by friction.
1799	The Royal Institution of Great Britain founded in London.
	Expiry of Watt's patents; introduction of high-pressure steam engines by Trevithick.
	Schelling's *System of Nature Philosophy [Naturphilosophie]*.
1800	Volta announces invention of electric battery.
	Herschel discovers infrared radiation.
1802	Dalton and Gay-Lussac publish independent experiments on the expansion of gases and establish Charles' law.
	Invention of fire-piston results in increased interest in adiabatic effects.
1804	Gay-Lussac and Biot hydrogen balloon ascents.
1807	Young describes *vis viva* as "energy."
1808	Dalton proposes atomic theory of chemical combination.
1813	Delaroche and Bérard measurements of specific heats of gases.

1815	Battle of Waterloo, fall of Napoleon.
	Dulong and Petit's papers on linear expansion of gases.
	Fresnel's first paper on diffraction.
1816	Rev. Stirling patents air-engine.
1819	Dulong and Petit's law of atomic heats.
1821	Herapath publishes version of kinetic theory of gases.
1822	Fourier's *Analytical theory of heat* on the mathematical theory of heat conduction.
1823	Poisson deduces formula for adiabatic expansion.
1824	Sadi Carnot's *Motive power of fire*.
	Liebig moves to Giessen, develops research laboratory.
1825	Stephenson constructs first steam locomotive railway.
1831	The British Association for the Advancement of Science is formed.
1833	Terms "scientist" and "physicist" coined by Whewell.
1834	Clapeyron publishes Carnot's ideas in more mathematical form.
1837	Coronation of Queen Victoria.
1840	Hess' law of heats of formation of chemical compounds.
1842	Mayer calculates equivalence of heat and work.
1843	Joule starts precision experiments on the mechanical equivalent of heat.
1845	Grove's *The correlation of physical force*.
1847	Helmholtz paper "On the Conservation of Force."
	Joule and Thomson meet at BAAS meeting in Oxford.
1848	"The Year of Revolutions" in Paris, Berlin, etc.
	Thomson constructs absolute temperature scale based on Carnot.
1850	Clausius' paper, "On the moving force of heat," reconciles Carnot's and Joule's ideas.
	Thomson confirms decrease of freezing point of ice under pressure.
1851	The Great Exhibition of the Works and Industry of All Nations in London.
1852	Thomson points out "universal tendency in nature to dissipation" and speaks of "conservation of mechanical energy."
1853	Rankine asserts "conservation of energy" in general.
1854	Thomson coins name "thermodynamics" for new science of heat and mechanics.
1857	Rankine states First and Second Laws of Thermodynamics.
	Clausius restates the basic kinetic theory of gases.
1859	Darwin publishes *On the origin of species*.
	Kirchhoff and Bunsen introduce spectral analysis.
1860	Maxwell develops kinetic theory of gases, allowing for a statistical distribution of molecular velocities.

1863	Andrews' experiments on non-ideal behavior of carbon dioxide gas.
1865	Clausius coins the name "entropy."
1866	Maxwell experiments on the viscosity of gases.
1867	Marx publishes *Das Kapital*.
1868	Mendeleev proposes periodic table of the chemical elements.
1870	The Franco-Prussian war and the Paris Commune.
1871	Maxwell publishes "Maxwell's demon" argument.
1872	Boltzmann devises H-function to explain Maxwell distribution.
	Maxwell publishes *Treatise on Electricity and Magnetism*, fully developing the view that light is an electromagnetic phenomenon.
1873–78	Gibbs' papers on chemical thermodynamics.
1873	Van der Waals' equation models non-ideal behavior of gases.
1874	Opening of Cavendish Laboratory, Cambridge.
1877	Boltzmann develops probabilistic model of entropy.
	Liquefaction of oxygen by Cailletet and Pictet.
1879	Stefan suggests Stefan-Boltzmann law.
1882	Helmholtz's paper on chemical thermodynamics.
1886	Langley measures distribution of black-body radiation.
1887	Van't Hoff treats osmotic pressure as extension of gas laws.
1896	Discovery of radioactivity by Becquerel.
	Commercial liquefaction of air by von Linde using Joule-Thomson effect.
	Boltzmann's *Lectures on gas theory*.
1898	Dewar liquefies hydrogen using Joule-Thomson effect.
1900	Planck proposes equation for black-body radiation implying quantization of energy.
1901	Death of Queen Victoria.
	Marconi achieves first transatlantic radio transmission.
1902	Gibbs publishes *Statistical mechanics*.
1905	Einstein's papers on special relativity and the photoelectric effect.
	Einstein explains Brownian motion in terms of molecular impacts.
1906	Nernst proposes "heat theorem," or Third Law of Thermodynamics.
1907	Einstein explains the specific heats of gases in quantum terms.
1908	First Model T car manufactured by the Ford Motor Co.
	Kamerlingh-Onnes liquefies helium.
	Perrin's experiments confirm Einstein's theory of Brownian motion.

1909–14   Development of the Haber-Bosch process for the manufac-
                   ture of ammonia from atmospheric nitrogen.
1911   First Solvay Conference on physics.
                   Kamerlingh-Onnes discovers superconductivity.
1914–18   The First World War.

GLOSSARY

Absolute temperature: See **Temperature, absolute.**

Absolute zero: As cold as it is possible to get, according to modern thermodynamics, although it is probably impossible ever to reach in practice. Equal to –273.15°C.

Action-at-a-Distance: The interaction of two objects that are separated in space with no known mediator of the interaction.

Actual energy: Early expression for **kinetic energy.**

Adiabatic: Of a process that takes place without the transfer of heat either into or out of a system.

Aether (ether): A hypothetical all-pervading substance assumed by most nineteenth-century scientists to be the medium for light and other **electromagnetic radiations.** Fell into disfavor after the publication of Einstein's theory of relativity.

Atmosphere: See **pressure.**

Atom: In ancient and early modern **natural philosophy,** an absolutely indivisible and fundamental constituent of matter. In modern science, any one of a hundred or so different particles from any one of which any particular chemical element is composed; no longer regarded as fundamentally indivisible. See **molecule; corpuscle.**

Black-body radiation: The characteristic distribution of **electromagnetic radiation** emitted by a "black body," i.e., a theoretical body that absorbs all the radiation that falls upon it.

Bolometer: An extremely sensitive instrument for measuring the intensity of radiant heat.

Boyle's law: That the **pressure** p and the volume V of a given mass of gas at a fixed temperature are inversely proportional to each other, or in other words that pV = constant. More-or-less true for **permanent gases** at normal temperatures and pressures. Also known as **Mariotte's law.**

British Thermal Unit (Btu): See **Heat energy.**

Brownian motion: The random motion of microscopic particles suspended in fluid, due to their random bombardment by the constituent molecules of the fluid.

Bushel: Traditional unit of volume or weight, approximately equal to about 8 gallons or 84 pounds.

Caloric: A weightless, indestructible fluid widely supposed by late-eighteenth- and early-nineteenth-century scientists to be responsible for the phenomena of heat.

Calorie: A measure of heat, defined as the amount of heat needed to raise the temperature of 1g of water by 1°C.

Calorific radiation: Old name for **radiant heat.**

Calorimeter: Apparatus for measuring the amount of heat released or absorbed by some process.

Carnot cycle: An idealized heat engine operating between a hot and a cold reservoir; the working substance (often taken to be an ideal gas) returns to exactly the same state at the end of the cycle as it was at the beginning.

Carnot function: According to Sadi Carnot, the (at the time) unknown function of temperature $C(\theta)$ which determined the **efficiency** of a heat engine; eventually turned out to be equal simply to the **absolute temperature,** T.

Charles' law: That the volume of a gas is a linear function of its temperature, i.e., for a given amount of gas at a constant pressure, the volume V at temperature θ is given by the formula $V = V_0(1 + \alpha\theta)$ where V_0 is the volume a 0°C and α is the (zero) coefficient of expansion, which is about 1/273 for any ideal gas.

Chemical Reaction: A process by which atoms of the same or different substances rearrange themselves to form a new substance. During this process, they either absorb heat or give it off.

Conduction (thermal): The process of heat transfer through a material, e.g., along an iron bar heated at one end.

Conductivity (thermal): A measure of the rate at which a particular substance conducts heat. Usually labeled K.

Convection: The process of heat transfer by the bulk transfer of material, e.g., by the hot air rising above a fire.

Corpuscle: In the works of **mechanical philosophers,** a very small particle; cf. **molecule.**

Critical temperature: See **vapor.**

Cryogenics: The science, practical and theoretical, of very low temperatures.

Diathermanous: Of a substance, such as quartz or rock-salt, that is transparent to **radiant heat;** glass, on the other hand, does not transmit radiant heat.

Diffusion (gaseous): The process by which one gas mixes with another.

Duty: A traditional British label either for the **power** of a machine (in horsepower), or for the **efficiency** of a machine, often in terms of the work done (e.g., volume of water pumped) for the consumption of a certain amount of fuel (e.g., a **bushel** of coal.)

Dynamical theory of heat: Any theory that proposes that heat is caused by the motion of the particles of matter.

Efficiency: The proportion of the energy input into a machine that can be extracted as useful work.

Elasticity: The readiness of a substance to be stretched or squashed, e.g., a coiled spring. Also applicable to air and other gases, which expand or contract as the external **pressure** changes.

Electrochemistry: Range of phenomena involving both electrical and chemical activity, e.g., an electric battery, electrolysis.

Electromagnetic radiation: A kind of radiation that has an enormous spectrum and variety, from x-rays (with very short wavelength and high frequency), through ultra-violet light, visible light, and **infrared radiation,** to microwaves and radio waves (with very long wavelength and low frequency).

Element: A fundamental component of a system.

Empirical: Based on sensory experience, observation and experiment. **Empiricism** is the philosophical theory that all sound knowledge must be so based.

Endothermic: Of a chemical reaction that absorbs heat. Cf. **exothermic.**

Energy: The universal currency of all activity in the material world, manifest in a variety of specific forms, e.g., kinetic (motion), gravitational, electrostatic, atomic. Measured in terms of **work** that can be performed, measured in joules.

Energy, free: The amount of energy that can be extracted from a system as useful work, and thus a measure of the stability of the system.

Energy, internal: The sum of the kinetic and potential energies of the atoms and molecules in a system; usual symbol U. A change in internal energy is not necessarily directly detectable as a change of temperature; see also **latent heat.**

Enlightenment: A philosophical movement in late-seventeenth- and eighteenth-century Europe that emphasized reason and the individual.

Entropy: A thermodynamic function that always increases in any spontaneous change in an isolated system; usual symbol S. Subsequently identified with the amount of disorder in a system.

Equation of state: An equation that relates the **pressure** p, volume V, and **temperature** of a substance; e.g., $pV = nRT$ for an **ideal gas.**

Equipartition of energy: The assumption that the energy of a system should be shared equally between the different forms that it might take, e.g., between translational, rotational, and vibrational kinetic energy for a diatomic molecule.

Ether: See **aether.** Also a highly volatile chemical (diethyl ether), once used as an anesthetic.

Exothermic: Of a chemical reaction that gives out heat. Cf. **endothermic.**

Function: A mathematical or physical relationship between two (or more) variables.

Heat energy: Energy involved in making bodies hotter or colder, changes of **state,** etc. Measured in units of the **calorie** (cal), i.e., the amount of energy required to raise the temperature of 1 g (gram) of water by 1°C, which equals 4.186 J. In Britain, traditionally measured in terms of the **British Thermal Unit** (Btu), i.e., the heat needed to raise the temperature of 1 lb of water by 1°F, which equals 252 cal. or 1054 J.

Ideal (or perfect) gas: A theoretical gas that obeys Boyle's and Charles' laws exactly. Its pressure p, volume V, and temperature T are therefore connected by the **equation of state** $pV = nRT$, where n is the amount of gas and R is the **universal gas constant,** which is 8.31 J/K.mol. Many **permanent gases** behave in an almost ideal fashion at normal temperatures and pressures.

Imponderable: Having no (detectable) weight, not "ponderable."

Infrared radiation: Electromagnetic radiation just below (in frequency) the red end of the spectrum of visible light; the means for the radiant transfer of heat energy from one body to another. See **radiant heat, calorific radiation.**

Isothermal: At a constant temperature.

Joule-Thomson (or Joule-Kelvin) effect: A slight change in temperature (usually a decrease) that gases exhibit when they expand without doing any external work; due to intermolecular forces.

Kinetic energy: The energy that a body has by virtue of its motion, either translational (i.e., from one place to another) or rotational or vibrational. The translational kinetic energy of a body of mass m and velocity v is $1/2 \, mv^2$.

Kinetic theory of gases: A theory that explains the physical properties of gases, such as pressure and temperature, in terms of the motions of the atoms or molecules of which the gas is formed. See **dynamical theory of heat.**

Latent heat: The heat that is absorbed (or given out) when a substance changes **state** from solid to liquid, or from liquid to gas (or vice versa).

Macrocosm: In medieval cosmology, the material world as a whole. Cf. **microcosm.**

Macroscopic: On a large scale; the bulk properties of things. Cf. **microscopic.**

Mariotte's law: See **Boyle's law.**

Material theory of heat: Any theory that proposes that heat is some kind of material substance; especially the **caloric** theory.

Maxwell (or Maxwell-Boltzmann) distribution: A formula that describes how the speeds of the constituent particles in a gas are distributed—both the spread around the average at a given temperature and the overall increase in speed as temperature increases.

Maxwell's equations: The set of equations, attributed to Maxwell, that describe the behavior of electric and magnetic fields, as well as the interactions of these fields with matter. Maxwell formulated these equations in 1865.

Mechanical equivalent (value) of heat: The amount of mechanical **work** required to generate a given amount of heat, and vice versa; 1 calorie = 4.186 J; 1 British thermal unit = 778 ft.lb. Usual symbol J.

Mechanical philosophy: A **natural philosophy** widely adopted in the seventeenth century, proposing that all natural phenomena could be explained in terms of the shape, size, and motions of fundamental material particles, whether **atoms** or other **corpuscles.** Especially associated with Descartes.

Medieval: That period of European history between ancient and **modern** times; roughly from the fall of the Roman Empire in fifth century A.D. until the **Renaissance** of the fifteenth century.

Microcosm: In medieval cosmology, the "lesser World" of man, i.e., the human body.

Microscopic: Either on a very small (maybe molecular) scale or, more literally, as seen through a microscope.

Modern: Historically, the period from the mid-fifteenth century onwards, i.e., from the end of the so-called Middle Ages. "Early modern" is usually taken to mean before the end of the eighteenth century; the French Revolution (1789–94) is often taken as a convenient watershed.

Molecule: In early natural philosophy, meant simply "small particle"; cf. **corpuscle.** In modern science, any particle consisting of two or more **atoms** joined together to make a chemical compound. Molecules with two atoms are "diatomic," those with three "triatomic," etc.

Natural philosophy: Traditionally, and until the early nineteenth century, the attempt to reach a comprehensive understanding of the fundamental causes of the natural world, generally envisaged as a divine creation. From the mid-nineteenth century displaced by a more specialized and secular "natural science." Not identical with the following.

***Naturphilosophie*:** A philosophy of nature prevalent in German literature, philosophy, and science from 1790 to 1830 or thereabouts. One of the central pillars of this intellectual movement was the belief in the unity of nature and its forces. Cf. **romanticism.**

Osmosis: The process by which a solvent will pass through a "semi-permeable" membrane from a more-concentrated to a less-concentrated solution.

Permanent gases: Gases, such as oxygen, nitrogen, and hydrogen, that cannot be liquefied at normal temperatures, no matter how much pressure is applied. See **Vapor.**

Phase: A distinct uniform part of a complex system, especially either solid, liquid, or gaseous. See **State.**

Phlogiston: A hypothetical substance supposed by some mid-eighteenth-century chemists to be responsible for combustibility, and thus to be released as flame during combustion. From Greek *phlox* = flame.

Potential energy: Energy belonging to a body because of force(s) acting on it; especially the gravitational potential energy of a body caused by Earth's gravity.

Power: The rate of working or otherwise transferring energy. In modern science measured in joules per second, or watts W. Traditionally, especially in Britain, measured in ft.lb. per minute or, in an engineering context, in horsepower (hp.); 1 horsepower equals 33,000 ft.lb./min. or 746 W.

Presbyterian: Of a Christian communion governed by a group of elders (presbyters) all of equal rank.

Pressure: Force applied per unit of area. In modern scientific terms, measured in newtons per square meter or pascal Pa. In engineering traditionally measured in pounds per square inch (lb/sq.in. or psi); 1 psi. = 6895 Pa. In the context of fluids and especially meteorology, older units are still used, e.g., 1 atmosphere (atm) = 1.013×10^5 Pa = 14.7 lb/sq.in. Air pressure is also measured in terms of the height of the column of mercury (chemical symbol, Hg) in a barometer; hence 1 atm. = 760 mm Hg = approx. 30 in Hg.

Quantum theory: Theory developed by Planck, Einstein, et al. from 1900 onwards. It proposes that electromagnetic radiation, and energy in general, only exists in discrete chunks or "quanta," of a size proportional to the frequency of the radiation. Subsequent quantum theory proposed conversely that all particles, such as electrons, have a wave-like character. Thus various paradoxes emerged, such as "wave-particle duality" and Heisenberg's "principle of indeterminacy."

Quasi-static: See **reversible.**

Radiant heat (or calorific radiation): The direct transfer of heat energy from one body to another by electromagnetic radiation, e.g., the heat detectable at some distance to the side of a fire. See **Infrared radiation.**

Reductionist: Of a theory that seeks to reduce one phenomenon (e.g., life or heat) to another (usually matter or mechanics). Cf. **vitalism.**

Renaissance: Rebirth, specifically the revival of interest in ancient Greek and Roman culture in fifteenth-century Europe. More generally applied to any similar cultural revival.

Reversible process: Idealized thermal process (e.g., the compression of a gas) proceeding through an infinite number of equilibrium states (hence also known as **quasi-static**) and therefore capable of being reversed (along exactly the same path) at any point by an infinitesimal change of variables. Also achieves theoretical maximum efficiency.

Romanticism: A late-eighteenth-century cultural movement that rejected much of the rationalism of the **Enlightenment.**

Scholastic: The intellectual style typical of the medieval universities or "schools," heavily based on the interpretation of traditional authorities such as Aristotle and early Christian theologians.

Scientist: Label coined in 1833 for the new breed of increasingly specialized, professional, and secular experts on natural science.

Specific heat capacity: The amount of heat needed by unit mass of a specific substance to increase its temperature by 1°C. ($Q = mc\theta$ where c = specific heat or heat capacity.) For a gas the specific heat may be measured either at constant volume (c_v) or at constant pressure (c_p); the ratio c_p/c_v is commonly labeled γ (gamma). By calculating the extra heat needed to compensate for the work done in expansion it is possible to show that, for an ideal gas, $c_p - c_v = R$.

State: The particular condition of something, especially whether it is solid, liquid, or gaseous.

State function: Banquet given by the Queen to honor a visiting head of state.

Statistical mechanics: A theory or model that treats the behavior of collections of objects in probabilistic or statistical terms, rather than attempting an exact, determinist mechanical treatment. Initially developed from the kinetic theory of gases in the late nineteenth century, and subsequently generalized, taking account of quantum theory, to explain the behavior of collections of other substances, such as photons and electrons.

Sublunary: In Aristotelian cosmology, the realm below the sphere of the moon, which is composed of the four elements and, unlike the heavens, subject to change and decay.

Temperament: A person's personality, especially as defined by its balance of the four humors, e.g., sanguine, choleric.

Temperature: A measure of how hot or cold something is; there are several more-or-less arbitrary temperature scales, e.g., the Celsius (or centigrade) scale, °C, and the Fahrenheit scale, °F. (In this text, such relative temperatures are designated as θ [theta].)

Temperature, absolute: The modern scientific thermodynamic temperature scale, measured in Kelvin K. This scale is independent of the properties of any particular substance. It starts from absolute zero. Usual symbol T.

Thermocouple: When the join between two wires of different metals is heated, an electric current may be generated.

Thermodynamics: The science of the interconversion of heat and work or, more generally, of all forms of energy.

Thermopile: A sensitive thermometer made from a large number of thermocouples; especially used for measuring radiant heat.

Torricellian: Of the vacuum produced at the top of a mercury barometer tube.

Transport properties: The physical properties of a gas that result from the physical interactions of its molecules, e.g., **diffusion, viscosity.**

Universal gas constant, R: See **Ideal gas.**

Vapor: A gas, such as carbon dioxide, that can be liquefied at room temperature by increased pressure. The temperature below which a gas is no longer **permanent** is its **critical temperature.** The critical temperature of carbon dioxide is 31.1°C.

Vis viva: The product of the mass of a body and the square of its speed; later identified with **kinetic energy.**

Viscosity: The extent to which a fluid (liquid or gas) resists the motion of bodies through it, or the friction between a fluid and a surface over which it is flowing. Oil is generally more viscous than water, for example.

Vitalism: The belief that living things are in some way intrinsically different from inanimate things and cannot be explained entirely in material terms.

Work: In modern science, the product of a force times the distance through which it acts; the modern unit, defined as a force of 1 newton x a distance of 1 meter, is the joule J; 1 J is roughly the amount of work needed to lift an apple up 1 meter. Usual symbol W. In Britain the traditional unit was the foot-pound (ft.lb.), which equals approximately 1.355J.

BIBLIOGRAPHY

Brush, Stephen G. 1965–72. *Kinetic Theory*. 3 vol. Oxford: Pergamon Press. Reissued, ed. N. S. Hall. London: Imperial College Press, 2003.

Volumes one and two contain a selection of important contributions to the theory of gas and heat from 1660 to 1900, including material by Boyle, Bernoulli, Mayer, Joule, Helmholtz, Clausius, Maxwell, Thomson, and Boltzmann.

Brush, Stephen G. 1976. *The Kind of Motion We Call Heat: A History of the Kinetic Theory of Gases in the 19th Century*. 2 vol. Amsterdam: North-Holland Publishing Company.

Detailed and fascinating study of key individuals and conceptual issues involved in the development of the kinetic theory in the nineteenth century.

Cannon, Susan Faye. 1978. *Science in Culture: The Early Victorian Period*. New York: Dawsons and Science History Publications.

Illuminating study of the emergence of the structures of modern science.

Cardwell, D. S. L. 1971. *From Watt to Clausius. The Rise of Thermodynamics in the Early Industrial Age*. London: Heinemann.

Study of the early development of thermodynamics, with particular reference to the influence of power technology.

Fox, Robert. 1971. *The Caloric Theory of Gases from Lavoisier to Regnault*. Oxford: Clarendon Press.

Definitive study of the rise and fall of the caloric theory from both conceptual and institutional points of view.

Gillispie, C. C., ed. 1970–80. *Dictionary of Scientific Biography [DSB]*. 16 vols.

Generally excellent introductions to the lives and works of both major and minor scientific figures.

Hankins, T. L. 1985. *Science and the Enlightenment.* Cambridge: Cambridge University Press.

> Clear, succinct survey of the conceptual development of science in the eighteenth century; includes a very useful bibliographical survey of the secondary literature.

Harman, P. M. 1982. *Energy, Force and Matter.* Cambridge: Cambridge University Press.

> Clear, succinct survey of the conceptual development of nineteenth-century physics; includes a very useful bibliographical survey of the secondary literature.

Hills, R. L. 1989. *Power from Steam. A History of the Stationary Steam Engine.* Cambridge: Cambridge University Press.

Holton, Gerald, and S. G. Brush. 2001. *Physics, the Human Adventure. From Copernicus to Einstein and Beyond.* New Brunswick, NJ: Rutgers University Press.

> An excellent historically structured introduction to modern physics.

Jungnickel, C., and R. McCormmach. 1986. *Intellectual Mastery of Nature. Theoretical Physics from Ohm to Einstein.* 2 vol. Chicago and London: Chicago University Press.

> Comprehensive study of the intellectual and institutional development of German physics in the nineteenth century.

Nye, Mary Jo, ed. 2003. *The Cambridge History of Science, vol. 5: The Modern Physical and Mathematical Sciences* (Cambridge: Cambridge University Press).

Olby, R. C. et al., ed. 1990. *Companion to the History of Modern Science.* London and New York: Routledge.

> Extensive collection of useful survey articles on many issues in the current history of modern science.

Smith, Crosbie. 1998. *The Science of Energy. A Cultural History of Energy Physics in Victorian Britain.* London: The Athlone Press.

> Fascinating investigation of construction of energy physics in nineteenth-century Britain, in the broadest personal, institutional, and cultural terms.

Weaver, J. H., ed. 1897. *The World of Physics. A Small Library of the Literature of Physics from Antiquity to the Present.* 3 vol. New York: Simon and Schuster.

> A useful anthology of original physics texts, including selections from the writings of Young, Rumford, Mayer, Joule, Helmholtz, Carnot, Clausius, Thomson, Gibbs, Boltzmann, and Planck.

PRIMARY SOURCES

Abbreviations of journal titles:

Phil. Mag. = *Philosophical Magazine*
Phil. Trans. = *Philosophical Transactions of the Royal Society of London*
Trans. or Proc. RSE = *Transactions* or *Proceedings of the Royal Society of Edinburgh*

Amontons, G. 1699. "Method of Substituting the Force of Fire for Horse and Man Power to Move Machines." *Histoire et mémoires de l'Académie Royale des Sciences.* Paris.

Anon. 1884. *Phil. Mag.* 18: 153–4.

Aubrey, J. 1972. *Aubrey's Brief Lives. Edited with the original manuscripts and with an introduction by Oliver Lawson Dick.* Harmondsworth: Penguin.

Bacon, Francis. 1620. *Novum Organum.* See Bacon, Francis. 1879. vol. 4.

———. 1627. *New Atlantis.* London: Printed by I[ohn] H[aviland and Augustine Mathewes] for William Lee at the Turks Head in Fleet-street, next to the Miter.

———. 1879. *The works of Francis Bacon.* Ed. J. Spedding et al. 14 vols. London: Longmans and Co. Reprint 1996. London: Routledge.

Bernoulli, D. 1968 [1738]. *Hydrodynamics.* Complete English transl. New York: Dover Publications. (Sections on kinetic theory quoted in Brush, 1965, pp. 57–65).

Berthollet, C. 1803. *Essai de statique chémique,* 2 vols. Paris.

Biot, J. B. 1816. *Traité de physique experimentale et mathematique.* 4 vols. Paris.

Black, Joseph. 1803. *Lectures on the Elements of Chemistry.* 2 vols. Edinburgh.

Blake, William. C. 1927 [c.1790]. *The Marriage of Heaven and Hell.* London: Dent.

Boerhaave, Hermann. 1741. *A New Method of Chemistry.* (English translation of 1732. *Elementa Chemiae.* Leiden). London.

Boltzmann, L. 1878 [1877]. "On the relationship between the second law of the mechanical theory of heat and the probability calculus." *Phil. Mag.* 6 (1878): 236. (Boltzmann, 1909, ii: 164–224).

———. 1964 [1896, 1898]. *Lectures on Gas Theory. (Vorlesungen uber Gastheorie).* Trans. S. Brush. Berkeley: University of California Press.

———. 1909. *Wissenschaftliche Abhandlungen [Scientific Papers].* 3 vol. Leipzig: J. A. Barth.

Boswell, J. 1934–50 (1791). *Life of Johnson.* Ed. G. Birkbeck-Hill. Oxford: Clarendon Press.

Boyle, R. 1660. *New Experiments Physico-Mechanicall, Touching the Spring of the Air, and its Effects; made, for the most part in a new pneumatical engine.* Oxford. (See Boyle, 1999, 1: 141–300).

———. 1662. *A Defence of the Doctrine Touching the Spring and Weight of the Air.* Oxford. (See Boyle, 1999, 3: 3–106).

———. 1665. *New Experiments and Observations Touching Cold, or an Experimental History of Cold, Begun.* London. (See Boyle, 1999, 4).

———. 1999. *The Works of Robert Boyle.* Ed. Michael Hunter and Edward B. Davis. 14 vols. London: Pickering and Chatto.

Carnot, Sadi. 1986 [1824]. *Reflexions on the Motive Power of Fire.* Trans. and ed. Robert Fox. Manchester: Manchester University Press.

Clapeyron, Emile. 1837 [1834]. "Memoir on the Motive Power of Heat." Ed. R. Taylor. *Scientific Memoirs,* 1: 347–76.

Clausius, R. 1851 [1850]. "On the moving force of heat, and the laws regarding the nature of heat itself which are deducible therefrom." *Phil. Mag. 2:* 1–21, 102–19. (Clausius, 1867, 14–69).

———. 1856 [1854]. "On a modified form of the second fundamental theorem in the mechanical theory of heat." *Phil. Mag.12:* 81–98. (Clausius, 1867, 111–135).

———. 1857. "On the nature of the motion which we call heat." *Phil. Mag. 14:* 108–207.

———. 1867. *The Mechanical Theory of Heat, with its Applications to the Steam Engine...*, edited by T. Archer Hirst. London: John Van Voorst.

Crawford, A. 1779. *Experiments and Observations on Animal Heat, and the Inflammation of Combustible Bodies.* London. (2nd ed., 1788. London).

Dalton, J. 1802. "On the expansion of elastic fluids by heat." *Memoirs [and proceedings] of the Literary and Philosophical Society of Manchester,* 5, part 2, 595–602.

———. 1808. *A New System of Chemical Philosophy.* Manchester. Reprint 1965. London: Peter Owen.

Darwin, C. 1892. *Charles Darwin: His Life told in an autobiographical Chapter, and in a selected series of his published letters.* Ed. Francis Darwin. London: J. Murray. Reprint 1950. New York: Schuman.

Delaroche, F., and Bérard, J. E. 1813. "Memoir on the determination of the specific heat of different gases." *Annales de chimie et de physique,* 85: 72–182.

Dulong, P. L., and A. T. Petit. 1818. "Researches on the measure of temperature and the laws of the communication of heat." *Annales de chimie et de physique,* 7. Paris.

Einstein, A. 1926. *Investigations on the theory of the Brownian Movement.* Trans. A. D. Cowper. London: Methuen. Reprint 1956. New York: Dover.

Fahrenheit, D. G. 1724. "Experimenta et observationes de Congelatione aquae in vacuo factae." *Phil. Trans., 33,* pp. 78–84.

Flammarion, C. 1917 [1893]. *La fin du monde.* Paris: Ernest Flammarion. (Trans. as *Omega: The Last Days of the World).*

Fourier, Jean-Baptiste Joseph. 1955 [1822]. *The Analytical Theory of Heat.* New York: Dover.

Frankel, Eugene. 1977. "J. B. Biot and the Mathematization of Experimental Physics in Napoleonic France." *Historical Studies in Physical Science,* 8 (1977), pp. 33–72.

Galilei, Galileo. 1957. *Discoveries and Opinions of Galileo. Translated with an Introduction and Notes by Stillman Drake.* New York: Doubleday Anchor.

Gibbs, J. W. 1902. *Elementary Principles in Statistical Mechanics.* New Haven, CT: Yale University Press.

———. 1906. *The scientific papers of J. Willard Gibbs.* 2 vols. London: Longmans, Green. Reprint 1961. New York: Dover.

———. 1948. *The Collected Works of J. Willard Gibbs.* New Haven: Yale University Press.

Groves, W. R. 1846. *On the correlation of physical forces.* London: S. Highley.

Haüy, R. J. 1806. *Traité élémentaire de physique.* 2nd ed., 2 vol., Paris.

Helmholtz, H. 1853 [1847]. "On the Conservation of Force." In J. Tyndall and W. Francis, eds. *Scientific Memoirs.* London: Taylor and Francis. 114–62.

———. 1856 [1854]. "On the interaction of natural forces." *Phil. Mag.* 11: 489–518.

Herapath, J. 1821. "A Mathematical Inquiry into the Causes, Laws and Principle Phenomena of Heat, Gases, Gravitation." *Annals of Philosophy* 1.

———. 1847. *Mathematical physics; or the mathematical principles of natural philosophy: with a development of the causes of heat, gaseous elasticity, gravitation, and other great phenomena of nature.* 2 vols. London: Whittaker and Co. (1972. Reprinted with an Introduction by S. G. Brush. New York: Johnson Reprint Corp.).

Herschel, J. 1865. "On the origin of force." *Fortnightly Review,* 1: 435–42.

Holtzmann, K.H.A. 1846 [1845]. "On the Heat and Elasticity of Gases and Vapours and on the Principles of the Theory of Steam Engines." In R. Taylor, ed., *Scientific Memoirs,* 4:189–217.

Jeans, Sir James. 1910. "On non-Newtonian mechanical systems, and Planck's theory of radiation," *Phil. Mag.* 20: 943–954.

Joule, J. P. 1839–40. "On electro-magnetic forces." *Ann. Electricity* 4: 474–81. (1884, i: 19–26).

———. 1840. "On the production of heat by voltaic electricity." *Proc. R. Soc.* (1884, i: 59–60).

———. 1843. "On the calorific effects of magneto-electricity, and on the mechanical value of heat." *Phil. Mag.* 23: 263–76, 347–55, 435–43. (1887, i: 123–59).

———. 1845. "On the changes of temperature produced by the rarefaction and condensation of air." *Phil. Mag.* 26: 369–83. (1884, i: 171–189).

———. 1846 [1843]. "On the heat evolved during the electrolysis of water." *Mem. Man. Lit. & Phil.* 7: 87–112. (1884, i: 109–23).

———. 1850. "On the mechanical equivalent of heat." *Phil. Trans.* 140: 61–82. (1884, i: 298–328).

———. 1857 [1851]. "Some remarks on heat, and the constitution of elastic fluids." *Phil. Mag.* 14: 211–16. (1884, i: 290–7).

———. 1884–87. *The Scientific Papers of James Prescott Joule.* 2 vols. London: The Physical Society. Reprint 1963. London: Dawsons.

Keats, J. 1973. *The Complete Poems.* Ed. J. Barnard. Harmondsworth; Baltimore: Penguin Education.

Kelland, P. 1837. *The Theory of Heat.* Cambridge: J. and J.J. Deighton.

Kelvin, Lord. See Thomson, W.

Langley, S. P. 1886. "Observations on invisible heat spectra...." *Phil. Mag.* 21: 394–409.

Laplace, P-S. 1799–1825. *Traité de Mécanique Celeste.* 5 vols. Paris.

Larmor, Sir Joseph. 1908. "Lord Kelvin." *Proc. Roy. Soc.* 81: iii-lxxvi.

Lavoisier, A. 1790 [1789]. *Elements of Chemistry in a new Systematic Order, containing all the modern discoveries.... Translated from the French, by Robert Kerr.* Edinburgh.

———. 1982 [1783]. *Memoir on Heat: translated with an introduction and notes by Henry Guerlac.* New York: Neale Watson Academic Publications Inc.

Leslie, J. 1804. An experimental inquiry into the nature and properties of heat. London.

Lodge, O. J. 1879. "An Attempt at a Systematic Classification of the Various Forms of Energy." *Phil. Mag.* 8: 277–86.

Martine, G. 1740. *Essays Medical and Philosophical.* London.

Maxwell, J. C. 1860. "Illustrations of the dynamical theories of gases." *Phil. Mag.* 21: 19–32; 20: 21–37.

———. 1866. "On the Viscosity or Internal Friction of Air and other Gases." *Phil. Trans* 156: 249ff.

———. 1867. "On the dynamical theory of gases." *Phil. Mag.* 35: 129–45, 185–217.

———. 1871. Theory of Heat. London: Longmans, Green.

———. 1876-7. "Hermann Ludwig Ferdinand Helmholtz." *Nature* 15: p. 389–91.

———. 1878. "Tait's 'Thermodynamics.'" *Nature* 17: 257–9, 278–80.

———. [1879]. "Does the Progress of Physical Science Tend to Give any Advantage to the Opinion of Necessity (or Determinism) over that of the Contingency of Events and the Freedom of the Will?" In Campbell and Garnett (1882).

———. 1890. *Scientific Papers,* 2 vols. Cambridge: Cambridge University Press.

———. 1908 [1876]. "On the equilibrium of heterogeneous substances." *Phil. Mag.* 16: 818–24.

———. 1990–2002. *The Scientific Letters and Papers of James Clerk Maxwell.* 3 vols. Ed. P. M. Harman. Cambridge: Cambridge University Press.

Mayer, J. R. 1842. "On the Forces of Inorganic Nature." *Annalen der Chemie und Pharmazie,* vol. 42, p. 233–40.

———. 1862 [1842]. "Remarks on the forces of inorganic Nature." *Phil. Mag.* 24: 371–7.

———. 1893. *Die Mechanik der Wärme.* Ed. J. J. Weyrausch. Stuttgart: Cotta.

Mohr, C.F. 1837. "Ansichten über die Natur der Wärme [Views on the Nature of Heat]," *Annalen der Chemie und Pharmazie* (ed. J. Liebig) 24: 141–7.

Nernst, W. 1909. *Theoretische Chemie.* 6th ed. Stuttgart: Enke.

Newcomb, S. 1862. *Proceedings of the American Academy of Arts and Sciences* 5, pp. 112ff.

Newton, Isaac. 1687. *Principia Mathematica Philosophiae Naturalis.* London.

———. 1701. "Scala graduum caloris [A scale of degrees of heat]." *Phil. Trans.* 22: 824–9.

———. 1934 [1729]. *Sir Isaac Newton's Mathematical Principles of Natural Philosophy and his System of the World. Principia,* 3rd ed. (1729). Trans. A. Motte. Revised by F. Cajori. Berkeley, CA: University of California Press.

Nietzsche, F. 1968 [1884–88]. *The Will to Power.* Trans. and ed. W. Kaufmann and R. J. Hollingdale. London: Weidenfeld and Nicolson.

Ostwald, W. 1906. *Individuality and immortality.* Boston: Houghton, Mifflin.

———. 1909. *Grundriss der allgemeinen Chemie,* 4th ed. Leipzig: Engelmann.

Perrin, J. 1923. *Atoms.* London: Constable.

Planck, Max. 1932. *Theory of Heat.* London: Macmillan.

Plato. 1965. *Timaeus.* Translated with an introduction by H. D. F. Lee. Harmondsworth, Penguin.

Power, H. 1663. *Experimental Philosophy: In Three Books; Containing New Experiments Microscopical, Mercurial, Magnetical.* London. Reprint 1966, with new introduction by M. B. Hall. New York: Johnson Reprint Corp.

Rankine, W.J.M. 1852. "On the reconcentration of the mechanical energy of the universe." *BAAS Report 22:* 12.

———. 1853. "On the general law of the transformation of energy." *Phil. Mag.* 5: 106–17.

———. 1860 [1857]. "Heat, Theory of the Mechanical Action of, or Thermodynamics." In Nichol, J. P., ed. *Cyclopaedia of the Physical Sciences.* 2nd ed. London: C. Griffin and co.

———. 1881. *Miscellaneous Scientific Papers.* Ed. W. J. Millar. London: Griffin.

Rayleigh, Lord [J. W. Strutt]. 1872. *Phil. Mag.* 4: 44.

———. 1875. "On the dissipation of energy." *Nature* 11: 454–5.

———. 1890. *Nature* 41: 26.

Rayleigh, Lord [R. J. Strutt]. 1924. *John William Strutt. Third Baron Rayleigh.* London: Edward Arnold.

Robsion, J. 1797. "The Steam Engine." *Encyclopaedia Britannica.* 3rd. ed. Edinburgh: Bell and Macfarquhar.

Rumford, Count. See Thomson, B.

Scheele, Carl Wilhelm. 1970 [1777]. *Chemical Treatise on Air and Fire.* Stockholm: Bokforlaget Rediviva.

Shelley, Mary. 1818. *Frankenstein, or, The modern Prometheus.* London: Printed for Lackington, Hughes, Harding, Mayor, & Jones.

Smeaton, J. 1759. "An experimental Enquiry concerning the natural Powers of Water and Wind to turn Mills, and other Machines, depending on a circular Motion." *Phil. Trans.* 51: 100ff.

Somerville, Mary. 1834. *On the Connexion of the Physical Sciences.* London: J. Murray.

Strutt, J. W. and R. J. See Rayleigh, Lord.

Tait, P. G. 1868. *Sketch of Thermodynamics.* Edinburgh: Edmonston and Douglas.

Thomson, B. 1798. "An Inquiry concerning the Source of the Heat which is Excited by Friction." *Phil. Trans.* 88: 80–102.

Thomson, W. (Lord Kelvin). 1848. "On an absolute thermometric scale, founded on Carnot's theory of the motive power of heat, and calculated from the results of Regnault's experiments on the pressure and latent heat of steam." *Phil. Mag.* 33: 313–17. (1882–1911, 1: 100–106).

———. 1849. "An account of Carnot's theory of the motive power of heat; with numerical results derived from Regnault's experiments on steam." *Transactions of the Royal Society of Edinburgh,* 16: 541–74.

———. 1852. "On a universal tendency in nature to the dissipation of mechanical energy." *Proc. RSE* 3: 139–42. (1882–1911, 1).

———. 1853 [1851]. "On the dynamical theory of heat; with numerical results deduced from Mr. Joule's 'Equivalent of a Thermal Unit' and Mr Regnault's 'Observations on Steam.'" *Phil. Mag.* 4: 88–21, 105–17, 168–76; *ibid.* 9: 523–31. (1882–1911, i: 174–232).

———. 1854. "Thermo-electric currents." *Trans. RSE* 21. (1882–1911, i: 232–291).

———. 1874. "The kinetic theory of the dissipation of energy." *Proc. RSE,* vol.8, pp. 325–34. (1882–1911, vol. 5, pp.11–20). (Brush, 1965, ii: 176–89).

———. 1881. "On the sources of energy available to man for the production of mechanical effect." *BAAS Rep. 51:* 513–18. (1889–94, 2: 433–52).

———. 1882–1911. *Mathematical and Physical Papers.* 6 vols. Cambridge: Cambridge University Press.

———. 1889–94. *Popular Lectures and Addresses.* 3 vols. London: Macmillan.

———. 1904. *Baltimore lectures on molecular dynamics and the wave theory of light.* London: Cambridge University Press.

Tyndall, John. [1874]. "The Belfast Address." In Basalla, George, et al. 1970. *Victorian Science.* New York: Anchor, pp. 435–78.

Waterston, J. J. 1893 [1846]. "On the physics of media that are composed of free and perfectly elastic molecules in a state of motion." (With introduction by Rayleigh). *Phil. Trans.* 183A: 5–77.

ADDITIONAL SECONDARY SOURCES

Abbreviations of journal titles:

Arch. Hist. Ex. Sci = *Archive for History of Exact Sciences*
BJHS = *British Journal for the History of Science*

Boas, Marie. 1962. *The Scientific Renaissance.* London: Collins.

Brush, S. G. 1978. *The Temperature of History. Phases of Science and Culture in the Nineteenth Century.* New York: Burt Franklin & Co.

———. 1981. *Statistical Physics and the Atomic Theory of Matter from Boyle and Newton to Laudan and Onsager.* Princeton, NJ: Princeton University Press.

Campbell, Lewis, and William Garnett. 1882. *The Life of James Clerk Maxwell.* London: Macmillan.

Cardwell, D. S. L. 1966. "Some factors in the early development of the concepts of power, work and energy." *BJHS* 3: 209–224.

———. 1989. *James Joule. A biography.* Manchester and New York: Manchester University Press.

———. 1994. *The Fontana History of Technology.* London: Fontana Press.

Conant, J. B. 1951. *Science and Common Sense.* New Haven: Yale University Press.

Crosland, M. 1967. *The Society of Arcueil.* London: Heinemann.

———. 1978. *Gay-Lussac. Scientist and Bourgeois.* Cambridge: Cambridge University Press.

Edser, E. 1923 [1899]. *Heat for advanced students.* London: Macmillan.

Fox, R. 1990. "Laplacian Physics." In Olby et al., ed., 1990, 278–294.

Garber, E., S. G. Brush, and C.W.F. Everitt, eds. 1986. *Maxwell on Molecules and Gases.* Cambridge, MA: MIT Press.

———. 1995. *Maxwell on Heat and Statistical Mechanics.* Bethelem, PA: Lehigh University Press.

Gillespie, C.C. "Carnot, Lazare." In Gillispie et al., ed. 1970-80. Vol 3: 70–79.

Grant, E., ed. 1974. A *Source Book in Medieval Science.* Cambridge, MA: Harvard University Press.

Hall, A. Rupert. 1963. *From Galileo to Newton.* London: Collins.

Klein, M. J. 1972. "Gibbs, Josiah Willard." In Gillispie et al., ed. 1970-80. Vol 5: 386–393.

Jamieson, A. 1900. *Elementary manual on steam and the steam engine.* 7th ed. London: Griffin.

Knott, C. G. 1911. *Life and scientific work of Peter Guthrie Tait.* Cambridge: Cambridge University Press.

Kuhn, T. S. 1977 (1959). "Energy conservation as an example of simultaneous discovery." Reprinted in Kuhn, ed. 1977. *The Essential Tension.* Chicago: Chicago University Press.

———. 1978. *Black-Body Problem and the Quantum Discontinuity, 1894–1912.* New York: Oxford University Press.

Leff, H. S., and A. F. Rex, eds. 1990. *Maxwell's Demon. Entropy, Information, Computing.* Bristol: Adam Hilger.

Lindemann, Mary. 1999. *Medicine and Society in Early Modern Europe.* Cambridge: Cambridge University Press.

McKie, D. and N. H. de V. Heathcote. 1935. *The Discovery Of Specific And Latent Heats.* London: Edward Arnold & Co.

Mendelsohn, Everett. 1964. *Heat and Life. The Development of the Theory of Animal Heat.* Cambridge, MA: Harvard University Press.

Mendelssohn, K. 1966. *The Quest for Absolute Zero: The Meaning of Low Temperature Physics.* London: Weidenfeld and Nicolson.

Middleton, W. E. Knowles. 1966. *A History of the Thermometer and Its Use in Meterology.* Baltimore: Johns Hopkins Press.

Muirhead, J. P. 1858. *The Life of James Watt.* London: Murray.

Myers, G. 1985–86. "Nineteenth-century popularizations of thermodynamics and the rhetoric of social prophecy." *Victorian Studies* 29: 35–66.

Nye, Mary Jo. 1972. *Molecular Reality.* New York: American Elsevier.

Pippard, A. B. 1957. *Elements of classical thermodynamics.* Cambridge: Cambridge University Press.

Preston, T. 1904. *Theory of Heat.* 2nd ed., revised by J. R. Cotter. London: Macmillan.

Schelar, V. M. 1966. "Thermochemistry and the third law of thermodynamics." *Chymia* 11: 99–124.

Schlipp, P. A., ed. 1949. *Albert Einstein Philosopher-Scientist.* New York: Library of Living Philosophers.

Sklar, L. 1995. *Physics and Chance: Philosophical Issues in the Foundations of Statistical Mechanics.* Cambridge: Cambridge University Press.

Smith, C. 1976. "William Thomson and the creation of thermodynamics: 1840–1855." *Arch. Hist. Ex. Sci.* 16: 231–88. Contains draft of Thomson 18531[851].

Smith, C. and M. N. Wise. 1989. *Energy and Empire. A biographical study of Lord Kelvin.* Cambridge: Cambridge University Press.

Talbot, G. R., and A. J. Pacey. 1966. "Some early kinetic theories of gases: Herapath and his predecessors." *BJHS* 3: 133–49.

Wear, A. 2000. *Knowledge & Practice in English Medicine, 1550–1680.* Cambridge, Cambridge University Press.

Zemansky, M. W. 1957. *Heat and Thermodynamics.* New York: McGraw-Hill Book Co.

INDEX

About the Author

CHRISTOPHER J. T. LEWIS teaches the history of science at the University of Cambridge. He studied natural science at Cambridge and the history of science at Imperial College, London. He has held research fellowships at the Warburg Institute, University of London, and the University of Padua in Italy. During the 1980s and 1990s, he worked as a tutor and lecturer in science and mathematics and the history of science, mathematics, and technology for the Open University.